TWELFTH EDITION

LABORATORY MANUAL ~~...RBOOK IN~~

MICROBIOLOGY

APPLICATIONS TO PATIENT CARE

Paul A. Granato, Ph.D., D(ABMM), F(AAM)

Professor Emeritus, Department of Pathology
SUNY Upstate Medical University
Director of Microbiology
Laboratory Alliance of Central New York
Syracuse, New York

Verna Morton, M.S., M.T. (ASCP)

Formerly Microbiology Section Head
Cox Health, Springfield, Missouri

Josephine A. Morello, Ph.D., D(ABMM), F(AAM)

Professor Emerita, Department of Pathology
Formerly Director of Hospital Laboratories and
Director of Clinical Microbiology Laboratories
The University of Chicago, Chicago, Illinois

Mc
Graw
Hill
Education

LABORATORY MANUAL AND WORKBOOK IN MICROBIOLOGY:
APPLICATIONS TO PATIENT CARE

Some ancillaries, including electronic and print components, may not be available to customers outside the
United States.

This book is printed on acid-free paper.

1 2 3 4 5 6 7 8 9 BKM 21 20 19 18

ISBN 978-1-260-09302-5
MHID 1-260-09302-6

Cover Image Credit: *Shutterstock / AuntSpray*

All credits appearing on page or at the end of the book are considered to be an extension of the copyright page.

The Internet addresses listed in the text were accurate at the time of publication. The inclusion of a website does
not indicate an endorsement by the authors or McGraw-Hill Education, and McGraw-Hill Education does not
guarantee the accuracy of the information presented at these sites.

Contents

Contents

PREFACE

To the Student:

You are about to begin the study of a fascinating group of organisms that are of immense importance in human health and disease. Although infectious diseases were once predicted to disappear with the widespread use of antimicrobial agents, the microbes have retaliated by means of their remarkable genetic versatility. Not only have some species acquired resistance to almost all existing antimicrobial agents but, in part due to social and political unrest, some old diseases have returned. New diseases, even some forms of cancer, are now known to be caused by microbes. Thus, whether your career leads you to direct patient contact as a nurse, or you pursue another patient care responsibility in a field of allied health, knowledge of microbiology is essential to perform your duties optimally and to protect both patients and yourself from acquisition and transfer of sometimes deadly pathogens.

In this context, this laboratory manual leads you through a series of exercises that allow you to learn not only basic techniques for working with microbes similar to those you will encounter in the clinical setting, but they will teach you how to practice safety precautions in the laboratory and hospital environment. Your journey through these experiments will be guided by your instructor, who will determine the time and materials available to perform each exercise. Most likely not every exercise will be performed in your course, but enough of them are available to provide you with a realistic idea of your responsibilities. You will learn what happens to a patient specimen after it is collected for microbiology laboratory analysis and how preliminary and completed laboratory reports are conveyed to the physician.

Several of the procedures you will perform have now been replaced by automated, immunological, or molecular methods, as will be described. However, the exercises in this manual will provide a basic understanding of microbes—their morphology, biochemical activity, immunological characteristics, and susceptibility to antimicrobial agents.

Before each laboratory session, your responsibility is to read through the introductory material of the exercise assigned and the procedures that you will follow, keeping in mind the Learning Objectives that precede each exercise. In this way, you will be able to understand the concepts introduced and complete the work in the time allotted. In some instances, you may work alone, or depending on the instructor and class size, you may work in pairs or groups. In either case, you should be familiar with all procedures and the results of each experiment. At the end of each exercise, you will be asked a series of questions. The question pages are set up so that they can be removed from the manual and handed in to the instructor without destroying the exercises. Be aware that not all answers to the questions are found in the exercise material but will require supplementary reading in your textbook and other literature or a search on the internet. With the rapid pace of medical knowledge, the internet has become a valuable resource to your continuing education, but only if you learn to recognize valid sites, such as those of the Centers for Disease Control and Prevention (www.cdc.gov); the National Library of Medicine (www.nlm.nih.gov, which includes PubMed Central); *Web* MD (www.webmd.com); and ClinLab Navigator (www.clinlabnavigator.com).

During our careers, we authors have been fascinated by the widespread prominence of microbes and their impact on health-related fields. We hope that your introduction to microbiology will increase your appreciation for their importance and need for respectful treatment.

To the Instructor:

This manual, now in its twelfth edition, maintains its original emphasis on the basic principles of diagnostic microbiology for students entering the allied health professions. The authors have emphasized the purposes and function of the clinical microbiology laboratory in the diagnosis of infectious diseases. The exercises illustrate as simply as possible the nature of laboratory procedures used for isolation and identification of infectious agents, as well as the principles of asepsis, disinfection, and sterilization. With the advent of automated, immunological, and molecular methods, formerly standard techniques for culture identification and antimicrobial susceptibility testing are being replaced by procedures that allow for more rapid test results. These methods are described in each exercise when appropriate, but we believe that the conventional methods practiced herein will help improve students' basic understanding of the morphology, biochemical activity, and immunological characteristics of important microbial pathogens.

Attention is also given to the role of the health professional in regard to appropriate collection of clinical specimens and the applications of aseptic and disinfectant techniques as they relate to patient care. In this way, the student receives the foundation needed to interpret patient-related information for the diagnosis and treatment of infectious diseases.

In this edition of the manual, each exercise has been carefully reviewed for accuracy and currency and revised when necessary to conform to changing practices in clinical laboratories. Two exercises from the eleventh edition, Preparing a Hanging Drop and Pour-Plate Technique, have been deleted because they play little role in current clinical laboratory practice. A description of the new multiplex syndrome panel testing technique using PCR has been added to the description of nucleic acid assays in a new exercise (Exercise 17). The MALDI-TOF instrument, which is gaining widespread use, is described in a new Exercise 18. For details, see What's New in the 12th Edition, which follows. Thought has been given to the time and resources available to instructors and students. Instructors may select among the exercises or parts of exercises they wish to perform, according to the focus of their courses.

A few exercises are primarily descriptive in nature because the equipment, supplies, or reagents needed to perform the relevant tests are beyond the resources available to most teaching laboratories. We considered that the students can benefit from knowing that such techniques are in regular use, even though they may not have the "hands-on" experience. If possible, it would be worthwhile for the instructor to arrange a field trip to a nearby clinical microbiology laboratory so that the students can view these diagnostic procedures firsthand.

Concern has been expressed that not all answers to the questions at the end of each exercise are found in the introductory or procedural material. This omission is purposeful to allow the students to think critically about the lessons and to explore alternative resources, as they will need to do throughout their careers. They should be directed to their textbooks, and in this computer age, to the internet, with caution about possible unreliable sources of information. Government sources such as the Centers for Disease Control and Prevention and the National Library of Medicine (see web addresses in To the Student) are usually accurate, but many other sources are available. The question section of each exercise is set up so that it can be removed from the manual and handed in to the instructor without destroying the exercises.

A complete *Instructor's Manual*, available online from McGraw-Hill Education, should greatly aid instructors' preparation for this course. Included are notes to instructors to help plan each exercise; formulae for preparation of reagents; preparation, storage, and sources of media; and suggested answers to questions in the manual. The URL and password for this site are available from your McGraw-Hill Education representative.

What's New in the 12th Edition

- All figures and colorplates have been carefully reviewed and changes made when necessary.
- In Exercise 3, the hanging drop preparation has been deleted because it is seldom used in clinical practice.
- Because former Exercise 6 (Special Stains) has been deleted from this edition, a brief discussion of capsules, flagella, and endospores has been included in Exercise 5 to orient the student to these bacterial structures.
- In Exercise 6, rather than having students prepare culture media, the media are prepared beforehand by the instructor. However, the students learn how to dispense them aseptically into Petri dishes or in tubes for sterilization.
- In Exercise 8, the pour-plate technique has been deleted because it is seldom used in clinical practice.
- In Exercise 10, conditions of pressure, temperature, and time currently used for steam pressure sterilization in most hospitals have been updated.

- In Exercise 12, a brief description of the problem of multidrug-resistant members of the *Entero-bacteriaceae* has been added. The information in table 12.1, Zone Diameter Interpretive Table, has been verified as conforming to the latest Performance Standards of the Clinical and Laboratory Standards Institute.
- Exercise 16 (formerly Exercise 17) is now limited to a discussion of the principles of antigen immunoassays. Although no experiments are performed in this exercise, the students are directed to experiments they will be performing in later exercises.
- The information on nucleic acid assays is now found in Exercise 17, along with a discussion of Multiplex Syndrome Panel Testing Using PCR Assays. This separation from the discussion of antigen immunoassays allows emphasis to be placed on the differences between the technologies and the use of each in the clinical microbiology laboratory.
- Exercise 18 is a new exercise discussing MALDI-TOF Mass Spectrometry for the Rapid Identification of Bacteria and Fungi. This instrument has come into widespread use in the clinical laboratory. The exercise is accompanied by figures describing the instrument and workflow.
- In Exercise 20, the use of PCR for the rapid identification of *Streptococcus pyogenes* in throat specimens is described. The availability of effective vaccines for pneumococcal pneumonia has been added. Figure 20.1 showing latex agglutination has been replaced.
- In Exercise 23, the advantage of molecular gene amplification methods in the detection of enteric pathogens is emphasized.
- In Exercise 26, the importance of the meningococcal vaccine in reducing the incidence of disease has been added.
- In Exercise 27, the importance of microarray assays for rapid identification of bacteria in blood cultures and the detection of antibiotic-resistance genes is described.
- Exercise 30 includes a discussion of the availability of nucleic acid amplification tests for organisms in this exercise. A section on the recent Ebola virus and Zika virus outbreaks has been added.
- Exercise 31 includes the use of nucleic acid and MALDI-TOF MS technologies for yeast identification.

Personalize Your Lab

Craft your teaching resources to match the way you teach! With McGraw-Hill Create™, www.mcgrawhillcreate.com, you can easily rearrange chapters, combine material from other content sources, and quickly upload content you have written like your course syllabus or teaching notes. Find the content you need in Create by searching through thousands of leading McGraw-Hill textbooks and lab manuals. Arrange your book to fit your teaching style. Create even allows you to personalize your book's appearance by selecting the cover and adding your name, school, and course information. Order a Create book and you'll receive a complimentary print review copy in 3–5 business days or a complimentary electronic review copy (eComp) via e-mail in minutes. Go to www.mcgrawhillcreate.com today and register to experience how McGraw-Hill Create empowers you to teach *your* students *your* way.

Acknowledgements

We acknowledge the role of McGraw-Hill Education, Lumina Datamatics, Inc., and Aptara, Inc. in the publication of this work. Their many courtesies—extended through Marija Magner, Senior Portfolio Manager, Darlene Schueller, Product Developer, Mary Jane Lampe, Content Product Manager, Shawntel Schmitt, Content Licensing Specialist, Cara Douglass-Graff, Senior Digital Content Developer; and Sarita Yadav, Senior Project Manager, Digital Publishing—have encouraged and guided this new edition. They have been primarily responsible for its production. For their skilled efforts and expert assistance, we thank Mathangi Anantharaman, Photo Researcher and Rajkishore Singh, Cover Designer.

P. A. G.

V. M.

J. A. M.

Orientation to the Microbiology Laboratory

Warning

Some of the laboratory experiments included in this text may be hazardous if you handle materials improperly or carry out procedures incorrectly. Safety precautions are necessary when you work with any microorganism and with Bunsen burners, chemicals, glass test tubes, hot water baths, sharp instruments, and similar materials. Your school may have specific regulations about safety procedures that your instructor will explain to you. If you have any problems with materials or procedures, please ask your instructor for help.

Safety Procedures and Precautions

The microbiology laboratory, whether in a classroom or a working diagnostic laboratory, is a place where cultures of microorganisms are handled and examined. This type of activity must be carried out with good aseptic technique in a thoroughly clean, well-organized workplace. In aseptic technique, all materials that are used have been sterilized to kill any microorganisms contained in or on them, and extreme care is taken not to introduce new organisms from the environment. Even if the microorganisms you are studying are not usually considered pathogenic (disease producing), *any* culture of *any* organism should be handled as if it were a potential pathogen. With current medical practices and procedures, many patients with lowered immune defenses survive longer than they did before. As a result, almost any microorganism can cause disease in them under the appropriate circumstances.

Each student must quickly learn and continuously practice aseptic laboratory technique. It is important to prevent contamination of your hands, hair, and clothing with culture material and also to protect your neighbors from such contamination. In addition, you must not contaminate your work with microorganisms from the environment. Once you learn the techniques for asepsis and proper disinfection in the laboratory, they apply to almost every phase of patient care, especially to the collection and handling of specimens that are critical if the laboratory is to make a diagnosis of infectious disease. These specimens should be handled as carefully as cultures so that they do not become sources of infection to others. An important problem in hospitals is the transmission of microorganisms between patients, especially by contaminated hands. Well-trained professionals, caring for the sick, should never be responsible for transmitting infection between patients. Appropriate attention to frequency and method of handwashing (scrubbing with soap for at least 30 seconds) or use of a hand sanitizing product, as described in Exercise 11, are critical for preventing these hospital-acquired infections (also known as nosocomial infections).

In general, all safety procedures and precautions followed in the microbiology laboratory are designed to:

1. *Restrict microorganisms present in specimens or cultures* to the containers in which they are collected, grown, or studied.
2. *Prevent environmental microorganisms* (normally present on hands, hair, clothing, laboratory benches, or in the air) from entering specimens or cultures and interfering with results of studies.

Hands and bench tops are kept clean with disinfectants, laboratory coats are worn, long hair is tied back, and working areas are kept clear of all unnecessary items. Containers used for specimen collection or culture material are presterilized and capped to prevent entry by unsterile air, and sterile tools are used for transferring specimens or cultures. *Nothing* is placed in the mouth.

Personal conduct in a microbiology laboratory should always be quiet and orderly. The instructor should be consulted promptly whenever problems arise. Any student with a fresh, unhealed cut, scratch, burn, or other injury on either hand should notify the instructor before beginning or continuing with the laboratory work. If you have a personal health problem and are in doubt about participating in the laboratory session, check with your instructor before beginning the work. *Careful attention to the principles of safety is required throughout any laboratory course in microbiology.*

General Laboratory Directions

1. Always read the assigned laboratory material *before* the start of the laboratory period. Pay particular attention to the Learning Objectives at the beginning of each exercise.
2. Before entering the laboratory, remove coats, jackets, and other outerwear. These should be left outside the laboratory, together with any backpacks, books, papers, or other items not needed for the work.
3. To be admitted to the laboratory, each student should wear a fresh, clean, knee-length laboratory coat.
4. At the start and end of each laboratory session, students should clean their assigned bench-top area with a disinfectant solution provided. That space should then be kept neat, clean, and uncluttered throughout each laboratory period.
5. Learn good personal habits from the beginning:
   Tie back long hair neatly, away from the shoulders.
   Do not wear jewelry to laboratory sessions.
   Keep fingers, pencils, and such objects out of your mouth.
   Do not smoke, eat, or drink in the laboratory.
   Do not lick labels with your tongue. Use tap water or preferably, self-sticking labels. Do not wander about the laboratory. Unnecessary activity can cause accidents, distract others, and promote contamination.
6. In the hospital and clinic setting, disposable gloves must be worn when patient care personnel draw blood from patients and when they collect and handle patient specimens. Once culture plates and tubes have been inoculated in the laboratory, however, gloves are not required to perform subsequent procedures.

7. Each student will need bibulous paper, lens paper, a china-marking pencil (or a black, waterproof marking pen) and a 100 mm ruler (purchased or provided). If Bunsen burners are used instead of bacterial incinerators, matches or flint strikers are also needed.

8. Keep a complete record of all your experiments, and answer all questions at the end of each exercise. Your completed work can be removed from the manual and submitted to the instructor for evaluation.

9. Discard all cultures and used glassware into the container labeled *CONTAMINATED* or *BIOHAZARD*. (This container will later be sterilized.) Plastic or other disposable items should be discarded separately from glassware in containers to be sterilized.
    Never place contaminated pipettes on the bench top.
    Never discard contaminated cultures, glassware, pipettes, tubes, or slides in the wastepaper basket or garbage can.
    Never discard contaminated liquids or liquid cultures in the sink.

10. If you are in doubt as to the correct procedure, double-check the manual. If doubt continues, consult your instructor. Avoid asking your neighbor for help with procedures.

11. If you should spill or drop a culture or if any type of accident occurs, *call the instructor immediately.* Place a paper towel over any spill and pour disinfectant over the towel. Let the disinfectant stand for 15 minutes, then clean the spill with fresh paper towels. Remember to discard the paper towels in the proper receptacle and wash your hands carefully.

12. Report any injury to your hands to the instructor either before the laboratory session begins or during the session.

13. Never remove specimens, cultures, or equipment from the laboratory under any circumstances.

14. Before leaving the laboratory, carefully wash and disinfect your hands. Arrange to launder your lab coat so that it will be fresh for the next session.

PART ONE

Basic Techniques of Microbiology

In the study of medical microbiology, we learn to recognize, isolate, and identify those microorganisms that are important in causing human infections and to differentiate them from harmless microorganisms that live in or on the body. We also learn methods to destroy them on animate and inanimate surfaces, such as on hands and in the hospital environment. In addition, we examine methods for guiding physicians in their choice of the most appropriate antimicrobial agent(s) for treating human infections.

In Part One, our study begins with the basic laboratory techniques that are needed to see and work with these microorganisms that are so important in health and disease. Thus, in Exercises 1 through 5, you will gain knowledge about the microscope, procedures for handling and examining cultures, and the staining techniques that enable us to learn about microbial morphology and certain characteristic structures. These steps are the first ones used in the laboratory identification of infectious agents. In Exercises 6 through 8, you will gain an understanding of the composition and use of culture media, which make it possible to grow and further study these microscopic organisms and confirm their role in specific infections. These same techniques are used throughout the study of microbiology, whether related to disease or to the surrounding environment (environmental microbiology).

The Microscope

Learning Objectives

After completing this exercise, students should be able to:

1. Describe the function of the following parts of the microscope:
   ocular lens
   objective lens
   iris diaphragm
   condenser
2. List three things they must do in order to put away their microscopes correctly.
3. Recall the difference between magnification and resolution.
4. State why oil is used on a slide to be examined with the oil-immersion objective.
5. Explain the advantage of parfocal objective lenses.

A good microscope is an essential tool for any microbiology laboratory. There are many kinds of microscopes, but the type most useful in diagnostic work is the *compound microscope.* By means of a series of lenses and a source of bright light, it magnifies and illuminates minute objects such as bacteria and other microorganisms that would otherwise be invisible to the eye. This type of microscope will be used throughout your laboratory course. As you gain experience using it, you will realize how precise it is and how valuable it is for studying microorganisms present in clinical specimens and in cultures. Even though you may not use a microscope in your profession, a firsthand knowledge of how to use it is important. Your laboratory experience with the microscope will give you a lasting impression of living forms that are too small to be seen unless they are highly magnified. As you learn about these "invisible" microorganisms, you should be better able to understand their role in transmission of infection.

Purpose	To study the compound microscope and learn
	A. Its important parts and their functions
	B. How to focus and use it to study microorganisms
	C. Its proper care and handling
Materials	An assigned microscope
	Lens paper
	Immersion oil
	A methylene-blue-stained smear of *Candida albicans* (a yeast of medical importance) or other microorganism (the fixed, stained smear will be provided by the instructor)

Instructions

A. Important Parts of the Compound Microscope and Their Functions

1. Look at the microscope assigned to you and compare it with the photograph in figure 1.1. Refer to table 1.1 for a summary of the parts of the microscope and their functions. Notice that the working parts are set into a sturdy frame consisting of a *base* for support and an *arm* for carrying it. (*Note:* When lifting and carrying the microscope, always use *both hands;* one to grasp the arm firmly, the other to support the base (fig. 1.2). *Never* lift it by the part that holds the lenses.)

2. Observe that a flat platform, or *stage* as it is called, extends between the upper lens system and the lower set of devices for providing light. The stage has a hole in the center that permits light from below to pass upward into the lenses above. The object to be viewed is positioned on the stage over this opening so that it is brightly illuminated from below. The stage has clips for holding a glass slide in place (do not attempt to place your slide on the stage yet). Note the *stage adjustment knobs* at the side of the stage in figure 1.1, which are used to move the slide in vertical and horizontal directions on the stage. This type of stage is referred to as a *mechanical stage.*

3. A built-in *illuminator* at the base is the source of light. Light is directed upward through the *Abbe condenser.* The condenser contains lenses that collect and concentrate the light, directing it upward through any object on the stage. It also has a shutter, or *iris diaphragm,* which can be used to adjust the amount of light admitted. A lever (sometimes a rotating knob) is provided on the condenser for operating the diaphragm.

   The condenser can be lowered or raised by an adjustment knob. Lowering the condenser decreases the amount of light that reaches the object. This is usually a disadvantage in microbiological work. It is best to keep the condenser fully raised and to adjust light intensity with the iris diaphragm.

4. The *rheostat control knob* at the base is used for adjusting the intensity of light emitted by the bulb.

5. Above the stage, attached to the arm, a tube holds the magnifying lenses through which the object is viewed. The lower end of the tube is fitted with a *rotating nosepiece* holding three or four *objective lenses.* As the nosepiece is rotated with the knurled ring, any one of the objectives can be brought into position above the stage opening. The upper end of the tube holds the *ocular lens,* or eyepiece (a monocular scope has one; a binocular scope permits viewing with both eyes through two oculars). Some microscopes are set up with an ocular micrometer. This is a disk that is marked with a ruled scale and is placed in one of the ocular-lens eyepieces. The ocular micrometer has been precisely measured (calibrated) for that particular microscope at different magnifications by using a stage micrometer. The stage micrometer is a microscope slide with a calibrated scale marked on its surface. By measuring the distance between the lines of the ocular and stage micrometers at each magnification, conversion factors are obtained that are used to determine the size of objects viewed with the ocular micrometer. Ocular micrometers are especially useful in identifying protozoa and other animal parasite forms, for example, helminth eggs, as seen in figure 32.3. Your instructor may have a microscope demonstration of an ocular micrometer.

6. Depending on the brand of microscope used, either the rotating nosepiece or the stage can be raised or lowered by *coarse* and *fine adjustment* knobs. These are located either above or below the stage. On some microscopes they are mounted as two separate knobs; on others they may be placed in tandem (see fig. 1.1) with the smaller fine adjustment extending from the larger coarse wheel. Locate the coarse adjustment on your microscope and rotate it gently, noting the upward or downward movement of the nosepiece or stage. The coarse adjustment is used to bring the objective into position over any object on the stage, *while looking at it from the side* to avoid striking the object and thus damaging the expensive objective lens (fig. 1.3). The fine adjustment knob moves the tube to such a slight degree that movement cannot be observed from the side. It is used when one is viewing the object through the lenses to make the small adjustments necessary for a sharp, clear image.

   Turn the adjustment knobs *slowly* and *gently,* as you pay attention to the relative positions of the objective and object. Avoid bringing the objective *down* (or the stage *up*) with the fine adjustment while viewing, because even this slight motion may force the lens against the object. Bring the lens safely in place first with the coarse knob; then, while looking through the ocular, turn the fine knob to adjust the lens until you have a clear view of the subject.

   Rotating the fine adjustment too far in either direction may cause it to jam. If this should happen, *never attempt to force it;* call the instructor. To avoid jamming, gently locate the two extremes to which the fine knob can be turned, then bring it back to the middle of its span and keep it within one turn of this central position. With practice, you will learn how to use the coarse and fine adjustment knobs in tandem to avoid damaging your slide preparations.

7. The *total magnification* achieved with the microscope depends on the combination of the *ocular* and *objective lens* used. Look at the ocular lens on your microscope. On most microscopes, you will see that it is marked "10×," meaning that it magnifies 10 times.

   Now look at the three objective lenses on the nosepiece. The short one is the *low-power* objective. Its metal shaft bears a "10×" mark, indicating that it gives tenfold magnification. When an object is viewed with the 10× objective combined with the 10× ocular, it is magnified 10 times 10, or ×100. Among your three objectives, this short one has the largest lens but the least magnifying power.

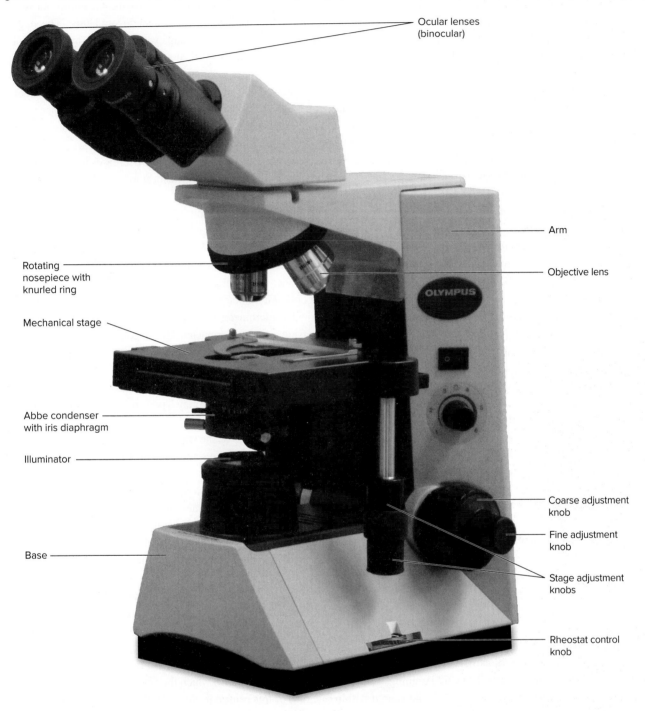

Ocular lenses
(binocular)

Arm

Rotating
nosepiece with
knurled ring

Objective lens

Mechanical stage

Abbe condenser
with iris diaphragm

Illuminator

Coarse adjustment
knob

Fine adjustment
knob

Base

Stage adjustment
knobs

Rheostat control
knob

Table 1.1   The Parts of the Microscope and Their Functions

Microscope Part	Function	Described in Instruction No.
Base	Serves as support; holds the illuminator	A1
Arm	Provides support between the tube and base; used also for support when carrying the microscope	A1
Stage	A platform for holding slides; has a center hole to permit light to enter from below into the lenses; the mechanical stage allows easy movement of slides	A2
Stage adjustment knobs	Used to move the slide on a mechanical stage in vertical and horizontal directions	A2
Illuminator	Provides the source of light; located in the base	A3
Abbe condenser	Collects and concentrates light upward through the object on stage	A3
Iris diaphragm	Adjusts the amount of light entering the condenser	A3
Rheostat control knob	Controls the amount of light emitted by the bulb	A4
Tube	Holds the ocular lenses at the top end and the rotating nosepiece with objective lenses at the lower end	A5
Rotating nosepiece	Holds the objective lenses, which are rotated by means of the knurled ring	A5
Objective lenses	Lenses of different power to magnify the object on the stage; usually 10x, 40x, and 100x power	A5
Ocular lenses	Lenses through which the object on the stage is viewed; may be monocular or binocular; usually provides additional 5x or 10x magnification	A5
Coarse adjustment knob	Depending on the type of microscope, allows either the stage or nosepiece to be *carefully* raised or lowered, respectively, to provide quick focus	A6
Fine adjustment knob	Once the object is in focus with the coarse adjustment knob, provides "fine tuning" of focus	A6

Figure 1.2   Proper handling of a microscope. Both hands are used when carrying this delicate instrument. ©*Josephine A. Morello*

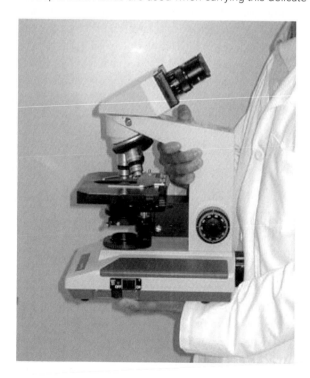

The Microscope

Figure 1.3 When adjusting the microscope, the technologist observes the objective carefully to prevent breaking the slide and damaging the objective lens of the microscope. This dual-view microscope has two microscope heads combined for simultaneous viewing and is extremely useful for teaching purposes. ©*Verna Morton*

The other two objectives look alike in length, but one is an intermediate objective, called the *high-power* (or *high-dry*) *objective*. It may or may not have a colored ring on it. What magnification number is stamped on it? _____ What is the total magnification to be obtained when it is used with the 10× ocular? _____

The third objective, which almost always has a colored ring, is called an *oil-immersion* objective. It has the smallest lens but gives the highest magnification of the three. (What is its magnifying number? _____ What total magnification will it provide together with the 10× ocular? _____) This objective is the most useful of the three for the microbiologist because its high magnification permits clear viewing of all but the smallest microorganisms (viruses require an electron microscope). As its name implies, this lens must be immersed in the drop of oil placed on the object to be viewed. The oil improves the *resolution* (the ability to distinguish detail) of the magnified image, providing a sharp image even though it is greatly enlarged. The function of the oil is to prevent any scattering of light rays passing through the object and to direct them straight upward through the lens.

Notice that the higher the magnification used, the more intense the light must be, but the amount of illumination needed is also determined by the density of the object. For example, more light is needed to view stained than unstained preparations.

8. The *focal length* of an objective is directly proportional to the diameter of its lens. You can see this by comparing your three objectives when each is positioned as close to the stage as the coarse adjustment permits. First place the low-power objective in vertical position and bring it down (or the stage up) with the coarse knob as far as it will go (gently!). When the object is in focus, the distance between the end of the objective, with its large lens, and the top of the cover glass is the focal length. Without moving the coarse adjustment, swing the high-power objective carefully into the vertical position, and note the much shorter focal length. Now, *with extreme caution,* bring the oil-immersion objective into place, making sure your microscope will permit this. If you think the lens will strike the stage or touch the condenser lens, *don't try it* until you have raised the nosepiece or lowered the stage (depending on your type of microscope) with the coarse adjustment. The focal length of the oil-immersion objective is between 1 and 2 mm, depending on the diameter of the lens it possesses (some are finer than others).

Never swing the oil-immersion objective into use position without checking to see that it will not make contact with the stage, the condenser, or the object being viewed. The oil lens alone is one of the most expensive and delicate parts of the microscope and must always be protected from scratching or other damage.

9. Take a piece of clean, soft *lens paper* and brush it lightly over the ocular and objective lenses and the top of the condenser. With subdued light coming through, look into the microscope. If you see specks of dust, rotate the ocular in its socket to see whether the dirt moves. If it does, it is on the ocular and should be wiped off more carefully. If you cannot solve the problem, call the instructor. *Never wipe the lenses with anything but clean, dry lens paper because solvents can damage the lenses.* Natural oil from eyelashes, mascara, or other eye makeup can soil the oculars badly and seriously interfere with microscopy. Eyeglasses may scratch or be scratched by the oculars. If they are available, protective eyecups placed on the oculars prevent these problems. If not, you must learn how to avoid soiling or damaging the ocular lens.

10. *If oculars or objectives must be removed from the microscope for any reason, only the instructor or other delegated person should remove them. Inexperienced hands can do irreparable damage to a precision instrument.*

11. Because students in other laboratory sections may also use your assigned microscope, *you should examine the microscope carefully at the beginning of each laboratory session. Report any new defects or damage to the instructor immediately.*

B. Microscopic Examination of a Slide Preparation

1. Now that you are familiar with the parts and mechanisms of the microscope, you are ready to learn how to focus and use it to study microorganisms. The stained smear provided for you is a preparation of a yeast (*Candida albicans*) that is large enough to be seen easily even with the low-power objective. With the higher objectives, you will see that it has some interesting structures of different sizes and shapes that can be readily located as you study the effect of increasing magnification. You are not expected to learn the morphology of the organism at this point (note that the instructor may substitute a different stained smear).

2. Place the stained slide securely on the mechanical stage fastened with the stage clips, making certain it cannot slip or move. Position it so that light coming up through the condenser passes through the center of the stained area.

3. Bring the low-power objective into vertical position and lower it (or raise the stage) as far as it will go with the coarse adjustment, observing from the side.

4. Look through the ocular. If you have a monocular scope, keep both eyes open (you will soon learn to ignore anything seen by the eye not looking into the scope). If you have a binocular scope, adjust the two oculars horizontally to the width between your eyes until you have a single, circular field of vision. Now bring the objective slowly upward or the stage downward with the coarse adjustment until you can see small, blue objects in the field. Make certain the condenser is fully raised, and adjust the light to comfortable brightness with the iris diaphragm.

5. Use the fine adjustment knob to get the image as sharp as possible. Now move the slide slowly around, up and down, back and forth. The low-power lens should give you an overview of the preparation and enable you to select an interesting area for closer observation at the next higher magnification. Record your observations in the circle labeled "Low-Power Objective" as described in step 9 on the next page.

6. When you have selected an area you wish to study further, swing the high-dry objective into place. If you are close to sharp focus, make your adjustments with the fine knob. If the slide is badly out of focus with the new objective in place, look at the body tube and adjust the lens close to, but not touching, the slide. Then, looking through the ocular, adjust the lens slowly, first with the coarse adjustment, then with the fine, until you have a sharp focus. Notice the difference in magnification of the structures you see with this objective as compared with the previous one. Record your observations as described in step 9.

7. Without moving the slide and changing the field you have now seen at two magnifications, wait for the instructor to demonstrate the use of the oil-immersion objective.

8. Move the high-dry lens a little to one side and place a drop of oil on the slide, directly over the stage opening. With your eyes on the oil-immersion objective, bring it carefully into position making certain it does not touch the stage or slide. Most microscopes are now *parfocal;* that is, the object remains in focus as you switch from one objective to another. In this case, the fine adjustment alone will bring the object into sharp focus. If not, while still looking at the objective, gently lower the nosepiece (or raise the stage) until the tip of the lens is immersed in the oil but is not in contact with the slide. Look through the ocular and very slowly focus upward with the fine adjustment. If you have trouble in finding the field or

getting a clear image, ask the instructor for help. When you have a sharp focus, observe the difference in magnification obtainable with this objective as compared with the other two. It is about $2\frac{1}{2}$ times greater than that provided by the high-power objective, and about 10 times more than that of the low-power lens.

9. Record your observations by drawing in each of the circles below several of the microbial structures you have seen, indicating their comparative size when viewed with each objective.

10. When you have finished your observations, lower the stage or raise the objective before removing the slide. Then remove the slide (taking care not to get oil on the high-dry lens). Gently clean the oil from the oil-immersion objective with a piece of dry lens paper.

Under each drawing, indicate the total magnification obtained by each objective combined with the ocular.

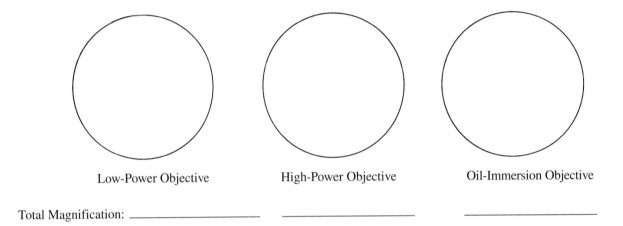

Low-Power Objective   High-Power Objective   Oil-Immersion Objective

Total Magnification: _____   _____   _____

C. Care and Handling of the Microscope

1. Always use both hands to carry the microscope—one holding the arm and one under the base (see fig. 1.2).
2. Before each use, examine the microscope carefully and report any unusual condition or damage.
3. Keep the oculars, objectives, and condenser lens clean. Use dry lens paper only.
4. At the end of each laboratory period in which the microscope is used, remove the slide from the stage, wipe away the oil on the oil-immersion objective, and place the low-power objective in vertical position.
5. Replace the dust cover, if available, and return the microscope to its box or assigned locker.

Table 1.2 suggests possible corrections to common problems encountered when using a microscope.

Table 1.2   Troubleshooting the Microscope

Problem	Possible Corrections
Insufficient light passing through ocular	Make certain the power cord is out of the way
	Raise the condenser
	Open the iris diaphragm
	Check the objective: is it locked in place?
Particles of dust or lint interfering with view of visual field	Wipe the ocular and the objective (*gently*) with clean lens paper
Moving particles in hazy visual field	Caused by bubbles in immersion oil; check the objective
	Make certain that the oil-immersion lens is in use, not the high-dry objective with oil on the slide
	Make certain the oil-immersion lens is in full contact with the oil.

Questions

1. Why is focal length important when using the oil-immersion objective?

2. Describe the best way to adjust the amount of light entering your specimen.

3. If a 5× ocular lens were used with your microscope, what maximum total magnification could be achieved?

4. When should the coarse and fine adjustment knobs be used?

5. If you see moving particles in a hazy visual field, what steps can you take to obtain a clearer image?

6. What would you observe if you forgot to use oil with the oil-immersion lens?

Handling and Examining Cultures

Learning Objectives

After completing this exercise, students should be able to:
1. Define a bacterial colony.
2. List five characteristics that are used to describe colonial morphology.
3. Explain why it is important to obtain pure cultures of bacteria.
4. Describe three forms of culture media and the advantages and disadvantages of each form.
5. List three aseptic techniques used in the laboratory to prevent contamination of the worker, the environment, and culture material.

Microscopic examination of microorganisms provides important information about their morphology but does not tell us much about their biological characteristics. Therefore, we need to observe microorganisms in *culture*. If we are to cultivate them successfully in the laboratory, we must provide them with suitable nutrients, such as protein components, carbohydrates, minerals, vitamins, and moisture in the right composition. This mixture is called a *culture medium* (plural, *media*). It may be prepared in liquid form, as a *broth,* or solidified with agar, a nonnutritive solidifying agent extracted from seaweed. *Agar media* may be used in tubes as a solid column (called a *deep*) or as *slants,* which have a greater surface area (see figs. 2.3 and 2.4). They are also commonly used in *Petri dishes* (named for the German bacteriologist who designed them), or *plates,* as they are often called.

Solid media are essential for isolating and separating bacteria growing together in a specimen collected from a patient; for example, urine or sputum. When a mixture of bacteria is streaked (spread) across the surface of an agar plate, it is diluted out so that single bacterial cells are deposited at certain areas on the plate. These single cells multiply at those sites until a visible aggregate called a *colony* is formed (see fig. 2.6). Each colony represents the growth of one bacterial species. A single, separated colony can be transferred to another medium, where it will grow as a *pure culture*. When certain patient specimens, such as respiratory and wound specimens, are inoculated onto agar plates, colonies of several different species often grow on them as a *mixed culture*. In order to study the properties of individual species without interference from other species, the microbiologist must work with single-colony, pure cultures. The practice of streaking plates to obtain pure cultures is critical in the hospital laboratory because it allows the microbiologist to determine how many types of bacteria are present, to identify those likely to be causing the patient's disease, and to test which antimicrobial agents will be effective for treatment. You will be learning the streaking technique to obtain pure cultures in Exercise 7.

The appearance of colonial growth on agar media can be very distinctive for individual species. Observation of the noticeable, gross features of colonies, such as color, density, consistency, surface texture, shape, and size, is very important (fig. 2.1a). These features, referred to as *colonial morphology,* can provide clues as to the identity of an organism. Final identification cannot be made by morphology alone, however.

In liquid media, some bacteria grow diffusely, producing uniform clouding known as turbidity, whereas others look very granular. Growth layered at the top, center, or bottom of a broth tube reveals something of the organisms' oxygen requirements. Sometimes colonial aggregates are formed and the bacterial growth appears as small puff balls floating in the broth. Observation of such features can also be helpful in recognizing types of organisms (fig. 2.1b).

Figure 2.1 Examples of bacterial growth patterns. (a) Some colonial characteristics on agar media. Characteristics of the colony edges may be distinctive for many bacterial species. The shapes and elevations shown in the two rows of sketches are not intended to be matched. (b) Some growth patterns in broth media.

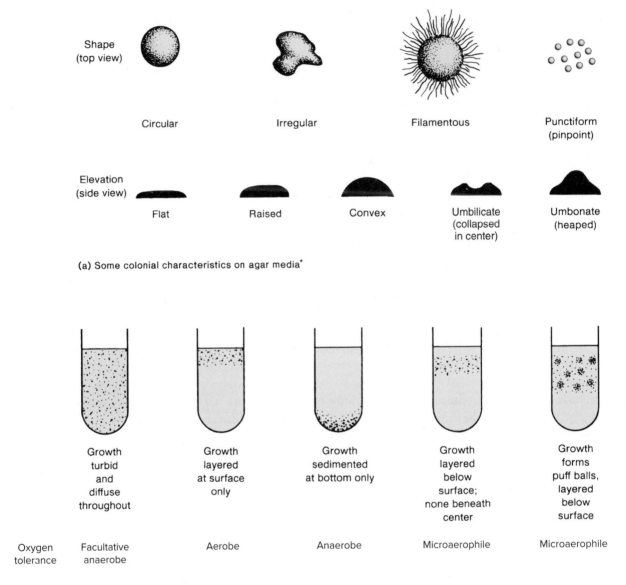

Shape (top view)

Circular   Irregular   Filamentous   Punctiform (pinpoint)

Elevation (side view)

Flat   Raised   Convex   Umbilicate (collapsed in center)   Umbonate (heaped)

(a) Some colonial characteristics on agar media*

Growth turbid and diffuse throughout

Growth layered at surface only

Growth sedimented at bottom only

Growth layered below surface; none beneath center

Growth forms puff balls, layered below surface

Oxygen tolerance   Facultative anaerobe   Aerobe   Anaerobe   Microaerophile   Microaerophile

(b) Some growth patterns in broth media

*Note: Shapes and elevations shown in this diagram are not intended to be matched.

You must learn how to handle cultures aseptically. The organisms must not be permitted to contaminate the worker or the environment, and the cultures must not be contaminated with extraneous organisms. In this exercise, you will use cultures containing environmental organisms or organisms of low pathogenic potential. Nonetheless, you should handle them carefully to avoid contaminating yourself and your neighbors. Also, if you contaminate the cultures, your results will be spoiled. Before you begin, reread the opening paragraphs of the section on Orientation to the Microbiology Laboratory, dealing with safety procedures and general laboratory directions (pages xi to xiii). In the clinical laboratory, current guidelines prohibit the use of open-flame burners, such as Bunsen burners. Instead, use of sterile disposable loops or needles or electric incinerators for sterilizing metal wire devices is advised. However, in some teaching

Basic Techniques of Microbiology

laboratories, Bunsen burners continue to be used. If so, special care must be taken to avoid contact of the flame with flammable materials as well as your hands, hair, and clothing.

Purpose	To make aseptic transfers of pure cultures and to examine them for important gross features
Materials	4 tubes of nutrient broth
	4 slants of nutrient agar
	One 24-hour slant culture of *Escherichia coli*
	One 24-hour slant culture of *Bacillus subtilis*
	One 24-hour slant culture of *Serratia marcescens* (pigmented)
	One 24-hour plate culture of *Serratia marcescens* (pigmented)
	Wire inoculating loop or disposable sterile inoculating loops
	Bunsen burner (and matches) or electric bacterial incinerator
	China-marking pencil or waterproof pen (or labels)
	A short ruler with millimeter markings
	Test tube rack

Procedures

A. Transfer of a Slant Culture to a Nutrient Broth

1. Before beginning the exercise, label all broth and agar tubes to be inoculated with your name and date, and label one set each with the name of the organisms to be studied. Be sure to place the label on an area of the tube that will not obscure the growth in the broth or on the slant after incubation. In practice, correct labeling of all specimen containers and culture plates and tubes that are used for patient material is a critical first step in avoiding diagnostic error.
2. The procedure will be demonstrated. Watch carefully and then do it yourself, following directions given.
3. Take up the inoculating loop by the handle and hold it as you would a pencil, loop down. Hold the wire in the flame of the Bunsen burner or in the bacterial incinerator until it glows red (fig. 2.2). The entire length of the wire that will enter the tube should be sterilized in this manner. Remove the loop and hold it steady a few moments until cool. *Do not wave it around, put it down, or touch it to anything.* Disposable loops that have been presterilized may be provided instead of a wire loop. When used to inoculate cultures, these should be handled in the same manner as described for sterilized wire loops.
4. Pick up the slant culture of *Escherichia coli* with your opposite hand. Still holding the loop like a pencil, but more horizontally, use the little finger of the loop hand to remove the closure (cotton plug, slip-on, or screw cap) of the culture tube. Keep your little finger curled around this closure when it is free—*do not place it on the table* (fig. 2.3).
5. Insert the loop into the open tube (holding both horizontally). Touch the loop (*not the handle!*) to the growth on the slant and pick up a *small* portion of the growth by gently touching the surface area. Do not dig the loop into the agar.
6. Withdraw the loop slowly and steadily, being careful not to touch it to the mouth of the tube. Keep it steady, *and do not touch it to anything* (it contains millions of bacterial cells!) while you replace the tube closure and put the tube back in the rack.
7. Still holding the loop steady in one hand, use the other hand to pick up a tube of sterile nutrient broth from the rack. Now remove the tube closure, as you did before, with the little finger of the loop hand (do not wave or jar the loop). Insert the loop into the tube and down into the broth. Gently rub the loop against the wall of the tube (do not agitate or splash the broth), making sure the liquid covers the area but does not touch the loop handle.
8. As you withdraw the loop, touch it to the inside wall of the tube (not the tube's mouth) to remove excess fluid from it. Pull it out without touching the tube wall again, replace the closure, and put the tube back in the rack.
9. Now carefully sterilize the loop unless you are using disposable sterile loops. In the latter case, discard the loop into a container of disinfectant or a biohazard bag (usually red or orange with the biohazard logo). A fresh disposable sterile loop should be used for each inoculation procedure. For wire loops, be sure all of the wire is sterilized, but do not burn the handle. When the wire has cooled, the loop can be placed on the bench top.
10. Repeat steps 3 through 9 with each of the other two slant cultures (*Bacillus subtilis* and *Serratia marcescens*).

Figure 2.2 Sterilizing the wire inoculating loop in the flame of a Bunsen burner (left) or a bacterial incinerator (right). *©Josephine A. Morello*

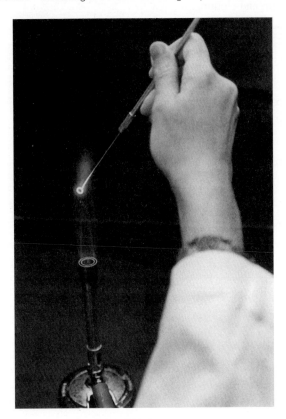 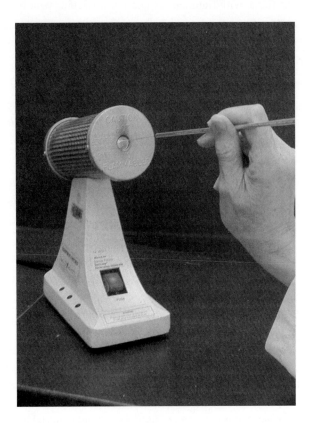

Figure 2.3 Inoculating a culture tube. Notice that the tube is held almost horizontally in one hand. Its cap is tucked in the little finger of the other hand, which holds the inoculating loop. *©Josephine A. Morello*

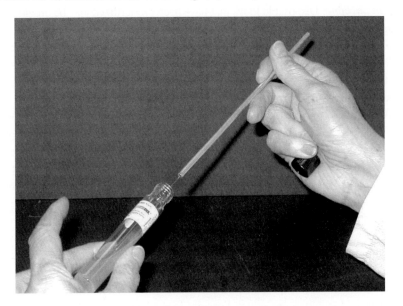

                                                              Basic Techniques of Microbiology

Figure 2.4   Streaking an agar slant with the loop.

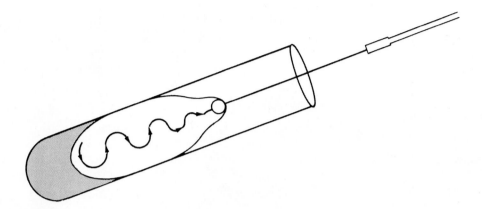

B. Transfer of a Slant Culture to a Nutrient Agar Slant

1. Start again with sterilizing the loop or use a disposable sterile loop.
2. Pick up the slant culture of *E. coli,* open it, and take up a small amount of growth on the sterile loop.
3. Recap the culture tube carefully and replace it in the rack. Pick up and open the sterile nutrient agar slant labeled *E. coli* (keep the loop containing the inoculum steady meantime).
4. Introduce the loop with the inoculum into the fresh tube of agar, and without touching any surface, pass it down the tube to the *deep* end of the slant. Streak the agar slant by lightly touching the loop to the surface of the agar, swishing it back and forth two or three times (don't dig up the agar), then zigzagging it upward to the top of the slant. Lift the loop from the agar surface and withdraw it from the tube without touching the tube surfaces (fig. 2.4).
5. Close and replace the inoculated tube in the rack; then sterilize or discard the loop as before.
6. Repeat steps 1 through 5 of procedure B with each of the other two slant cultures provided (*B. subtilis* and *S. marcescens*).

C. Transfer of a Single Bacterial Colony on a Plate Culture to a Nutrient Broth and a Nutrient Agar Slant

1. Start again with sterilizing the loop or use a disposable sterile loop. A loop that has been sterilized by heat must always be cooled by holding it steady for a few minutes before touching any bacterial growth.
2. Hold the sterile loop in one hand and with the other hand turn the assigned plate culture of *S. marcescens* so that it is positioned with the bottom (smaller) part of the dish up. The bottom contains the inoculated agar. Lift this part of the dish with your free hand (fig. 2.5) and turn it so that you can clearly see isolated colonies of *S. marcescens* growing on the surface of the plated agar.
3. With the sterile loop, touch the *surface* of one isolated bacterial colony (fig. 2.6). Withdraw the loop and replace the bottom part of the dish into the inverted lid lying open on the table.
4. Now inoculate a labeled sterile nutrient broth as in procedure A, steps 7 through 9.
5. With a sterilized loop, open the plate, pick another colony, close the plate, and inoculate a sterile agar slant as in procedure B, steps 4 and 5.

D. Incubation of Freshly Inoculated Cultures

1. Make certain all the broths (4) and slants (4) you have inoculated are properly and fully labeled.
2. Place your transferred cultures in an assigned rack in the incubator. The incubator temperature should be 35° to 37°C.

   Record your reading of the incubator thermometer here. _____

Figure 2.5 Opening a Petri dish culture. The bottom with the agar is lifted out of the lid, and the lid is left lying face up on the bench.
©Josephine A. Morello

Figure 2.6 Selecting an isolated bacterial colony from a plate culture surface. The plate has been streaked so that single colonies have grown in well-separated positions and can easily be picked up. *©Josephine A. Morello*

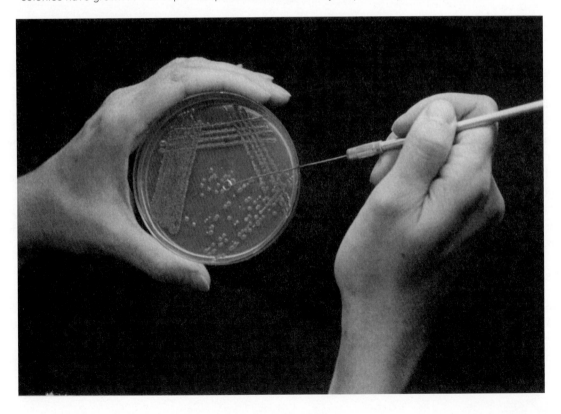

Basic Techniques of Microbiology

E. Examination of Culture Growth

1. When you have finished making the culture transfers as directed, take a few minutes to look closely at the grown cultures with which you have been working. In the Results section of this exercise, there are blank forms in which you can record information as to the appearance of these cultures, specifically, some or all of the following: *size of colonies* (in mm), *color, density* (translucent? opaque?), *consistency* (creamy? dry? flaky?), *surface texture* (smooth? rough?), and *shape of colony* (margin even or serrated? flat? heaped?).

2. The cultures you have made should be incubated for 18 to 24 hours before you examine them, to allow time for sufficient bacterial growth to appear. After this time, record their appearance in broth or on slants, using the blank forms in the Results section. Provide *all* the information the forms require, as in procedure E.1.

Results

Record your observations of all cultures in the tables or diagram. Consult section E.1 and figure 2.1 (Examples of bacterial growth patterns) for appropriate descriptive terms.

1. Slant cultures from which you made your inoculations.

Name of Organism	Appearance on Slants		
	Color	Density	Consistency
Escherichia coli			
Bacillus subtilis			
Serratia marcescens			

2. Colonies on plate culture of *S. marcescens*.

Size (mm)*	Color	Density	Consistency	Colony Shape; Surface Texture

*With your ruler, measure the diameter of the average colony on the plate culture by placing the ruler on the *bottom* of the plate. Hold plate and ruler against the light to make your readings.

3. The slant cultures you inoculated at the previous session.

Name of Organism	Appearance on Slants		
	Color	Density	Consistency
Escherichia coli			
Bacillus subtilis			
Serratia marcescens			
Serratia marcescens*			

*Inoculated from culture plate

If you have made successful transfers and achieved pure cultures, the morphology of your cultures should match that of the ones you were assigned. If not, explain why this might have happened and how it could have been prevented.

4. Refer to the bottom portion of figure 2.1 and shade in the type of growth you observed in your broth cultures.

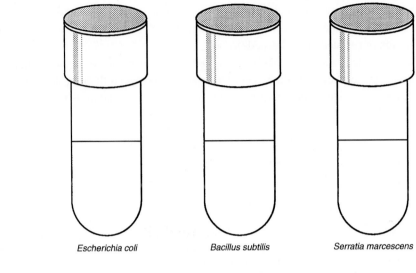

Escherichia coli   Bacillus subtilis   Serratia marcescens

Most Likely Oxygen Tolerance ——————————   ——————————   ——————————

Basic Techniques of Microbiology

Questions

1. How would you determine whether culture media given to you are sterile before you use them?

2. What are the signs of growth in a liquid medium?

3. What is the purpose of wiping the laboratory bench top with disinfectant before you begin to handle cultures?

4. Why can a single colony on a plate be used to start a pure culture?

5. Why is it important not to contaminate a pure culture?

6. What is meant by the term *colonial morphology*?

7. Why should long hair be tied back when one is working in a microbiology laboratory? Can you think of an actual patient-care situation that would call for its control for the same reason?

8. Name at least two kinds of solutions that may be administered to patients by intravenous injection and therefore must be sterile. How would you know if they were not sterile?

Simple Stains and Wet-Mount Preparations

Learning Objectives

After completing this exercise, students should be able to:
1. Explain why bacteria stain well with basic dyes.
2. Understand the purpose of fixing a slide that is to be stained.
3. List and define the three basic shapes of bacteria.
4. Compare and contrast true motility with Brownian movement.
5. Recall the level of light intensity that will make it easier to find organisms in a wet mount.

Now that you know some basic tools and methods used in microbiology, we shall learn how to make preparations to study microbial morphology under the microscope in both stained and natural states. Bacteria are so small and have so little substance that they tend to be transparent, even when magnified. In the first part of this exercise, you will learn how to make a simple stained preparation to visualize bacteria. Many sophisticated ways of doing this are known, but the simplest is to smear out a bacterial suspension on a glass slide, "fix" the organisms to the slide, then stain them with a visible dye (Robert Koch and his coworkers first thought of this more than 100 years ago).

Principles of Bacterial Stains and Morphology

The best bacterial stains are *aniline dyes* (synthetic organic dyes made from coal-tar products). When they are used directly on fixed bacterial smears, the shapes of bacterial bodies are clearly seen. These dyes react in either an *acidic, basic,* or *neutral* manner. Acidic and basic stains are used primarily in bacteriologic work. The free ions of *acidic* dyes are anions (negatively charged) that combine with cations of a base in the stained cell to form a salt. *Basic* dyes possess cations (positively charged) that combine with an acid in the stained material to form a salt. Bacterial cells are rich in RNA (contained in their abundant ribosomes) and therefore stain very well with basic dyes. *Neutral* stains are made by combining acidic and basic dyes. They are most useful for staining complex cells of higher forms because they permit differentiation of interior structures, some of which are basic, some acidic. Cells and structures that stain with basic dyes are said to be *basophilic.* Those that stain with acid dyes are termed *acidophilic.*

Stained bacteria can be measured for size and are classified by their shapes and groupings. Bacteria are so small that their size is most conveniently expressed in *micrometers* (symbol μm). A micrometer is a thousandth part of a millimeter, and 1/10,000 of a centimeter, or 1/25,400 of an inch. Bacteria vary in length and diameter, the smallest being about 0.5 to 1 μm long and approximately 0.5 μm in diameter, whereas the largest filamentous forms may be as long as 100 μm. Most of those you will see in this course are at the small end of the scale, measuring about 1 to 3 μm in length. Small as they are in reality, their images should loom large in your mind as the agents of infection in patients for whom you will be caring.

Figure 3.1   Basic shapes and arrangements of bacteria.

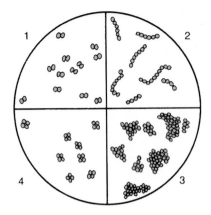
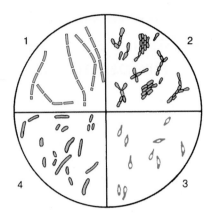
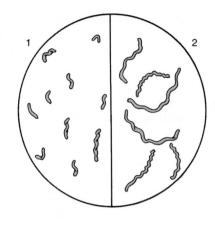

Cocci

1. Diplococci (pairs)
2. Streptococci (chains)
3. Staphylococci (grapelike clusters)
4. Tetrads (packets of four)

Bacilli (rods)

1. Streptobacilli (chains)
2. Palisades (V, X, Y forms, clubbing)
3. Endospore-forming (small, round hollow bodies in cell)
4. Pleomorphism (varying widths and lengths)

Spirals

1. Spirilla (short, curved, or spiral forms with rigid bodies)
2. Spirochetes (long, tightly or loosely coiled forms with flexible bodies)

Bacteria have rigid cell walls and maintain a constant shape that is used as a basis for their classification. Bacteria have three basic shapes: *spherical* (round), *rod shaped,* or *spiraled* (fig. 3.1). A round bacterium is called a *coccus* (plural, *cocci*). A rod-shaped organism is called a *bacillus* (plural, *bacilli*) or simply a *rod.* A spiraled bacterium with at least two or three curves in its body is called a *spirillum* (plural, *spirilla*). Long, sinuous organisms with many loose or tight coils are called *spirochetes.*

Individual bacterial genera or species are often characterized by the cell patterns they form when grouping together as they multiply. Cocci may occur in pairs (*diplococci*), chains (*streptococci*), clusters (*staphylococci*), or packets of four (*tetrads*), and are seldom found singly.

Rod-shaped bacteria (bacilli) generally occur as individual cells, but they may appear as end-to-end pairs (*diplobacilli*) or line up in chains (*streptobacilli*). Some species tend to *palisade,* that is, line up in bundles of parallel bacilli; others may form V, X, or Y figures as they divide and split. Some may show great variation in their size and length (known as pleomorphism).

Spiraled bacteria occur singly and usually do not form group patterns. Examine **colorplates 1–8** to see representative examples of bacterial morphology.

Viewing Bacteria in the Living State

In the second part of this exercise, you will study living organisms under the microscope. The simplest method for examining living microorganisms is to suspend them in a fluid (water, saline, or broth) and prepare a simple "wet mount." Microscopic study of such a wet preparation provides useful information, primarily whether or not an organism is motile.

True, independent motility of bacteria depends on their possession of flagella with which they can propel themselves with progressive, directional locomotion (often quite rapidly).

When viewing a wet mount, you must distinguish this kind of active motion from the vibratory movement of organisms or other particles suspended in a fluid. The latter type of motion is called *Brownian movement* and is caused by the continuous, rapid oscillation of molecules in the fluid. Small particles of any kind, including bacteria (whether motile or not), are constantly bombarded by the vibration of the fluid molecules, and so are bobbed up and down, back and forth. Such movement is irregular and nondirectional and does not cause nonmotile organisms to change position with respect to other objects around them.

You must be careful not to mistake movement caused by currents in a liquid for true motility. If a wet mount contains bubbles, air currents set up reacting fluid currents, and you will see organisms streaming along on a tide. A wet mount almost always has flowing currents.

EXPERIMENT **3.1**   **Preparing a Simple Stain**

Purpose	To learn the value of simple stains in studying basic microbial morphology
Materials	24-hour agar culture of *Staphylococcus epidermidis*
	24-hour agar culture of *Bacillus subtilis*
	24-hour agar culture of *Escherichia coli*
	Distilled water
	Prepared stained smear of a spiraled organism
	Methylene blue
	Absolute methanol (if used for fixation)
	Safranin
	Sterile toothpicks
	Slides
	Staining rack
	China-marking pencil or permanent marking pen
	Bibulous paper

Procedures

1. Slides for microscopic smears must always be sparkling clean. They may be stored or dipped in alcohol and polished clean (free of grease) with a tissue or soft cloth.
2. Take three clean slides and with your marking pencil or pen make a circle (about 1½ cm in diameter) in the center. At one end of the slide write the initials of one of the three assigned organisms (your three slides should read Se, Bs, and Ec, respectively). If slides with one frosted end are used, pencil marks on the frosted area will remain throughout the staining procedure.
3. Turn the slides over so that the unmarked side is up. (Except for slides with a frosted end, when slides are to be stained, pen or pencil markings should always be placed on the underside so that the mark will not smear, wash off, or run into the smear itself.)
4. With your inoculating loop, place a loopful of distilled water in the ringed area of the slide. Using proper aseptic transfer techniques, mix a **small** amount of bacteria in the water and spread it out. Be certain the smear is only *lightly* turbid. Repeat this step until smears of all three organisms have been made.
5. Allow the smears to air dry. You should be able to see a thin white film on each slide. If not, add another loopful of water and more bacteria, as in step 4.
6. If you are using a bacterial incinerator, heat-fix the smears by pressing the slides against the face of the incinerator for 2 seconds. Otherwise, heat-fix the smears by passing the slides rapidly through the Bunsen flame three times so that the smears will not wash off. Alternatively, the smears may be heat fixed on a slide warmer or fixed without heat by placing

the slides on a staining rack and flooding them with absolute methanol. Allow the slides to sit for 1 minute, then drain off the alcohol and air dry them completely.

7. Place the slides on a staining rack and flood them with methylene blue. Leave the stain on for 3 minutes.
8. Wash each slide gently with distilled water, drain off excess water, blot (do not rub) with bibulous paper, and let the slides dry completely in air.
9. While the slides are drying, take two more clean slides and draw a circle on the bottom with your marking pencil or pen.
10. Place a loopful of distilled water over the circle on each slide.
11. With the flat end of a toothpick, scrape some material from the surface of your teeth and around the gums. Emulsify the material in the drop of water on one slide. Repeat this procedure on the other slide.
12. Allow both slides to dry in air; then fix them with heat or methanol. Stain one slide with methylene blue for 3 minutes and the other with safranin for 3 minutes.
13. Wash, drain, and dry the slides as in step 8.
14. Examine all slides, including the prepared stained smear assigned to you, with all three microscope objectives. Record your results in the table.

Results

Organism in Broth Culture	Stain Used	Color	Coccus, Rod, or Spiral	Cell Grouping	Prepare a Diagram
Staphylococcus epidermidis					
Bacillus subtilis					
Escherichia coli					
Prepared smear of spiraled organism					

Draw the organisms you saw in the scraping from your teeth.

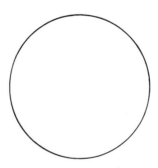

Describe the results you obtained with the two stains used. Which provided the sharpest view?

Methylene blue _____

Safranin _____

Basic Techniques of Microbiology

EXPERIMENT **3.2**   **Preparing a Wet Mount**

Purpose	To observe bacteria in a simple wet mount and determine their motility
Materials	24-hour broth culture of *Proteus vulgaris* mixed with a light suspension of yeast cells
	24-hour broth culture of *Staphylococcus epidermidis* mixed with a light suspension of yeast cells
	2 microscope slides
	Several clean cover glasses
	Capillary pipettes and pipette bulbs
	China-marking pencil or permanent marking pen

Procedures

1. Using a pipette bulb, aspirate a small amount of the *Proteus* culture with a capillary pipette and place a small drop on a clean microscope slide (fig. 3.2, step 1).
2. Take a cover glass and clean it thoroughly, making certain it is free of grease. It may be dipped in alcohol and polished dry with tissue, or washed in soap and water, rinsed completely, and wiped dry. Place the cover glass over the drop, trying to avoid bubble formation (fig. 3.2, step 2). The fluid should not leak out from under the edges of the cover glass. If it does, wait until the fluid around the edges dries before examining the preparation.

Figure 3.2   Wet-mount preparation.

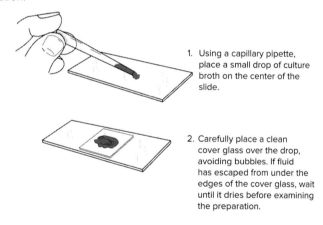

1. Using a capillary pipette, place a small drop of culture broth on the center of the slide.

2. Carefully place a clean cover glass over the drop, avoiding bubbles. If fluid has escaped from under the edges of the cover glass, wait until it dries before examining the preparation.

3. Place the slide on the microscope stage, cover glass up. Start your examination with the low-power objective to find the focus. It is helpful to focus first on one edge of the cover glass, which will appear as a dark line. The light should be reduced with the iris diaphragm, and if necessary, by lowering the condenser. If you have trouble with the focus, ask the instructor for help.
4. Continue your examination with the high-dry objective.
5. Make a wet-mount preparation of the *Staphylococcus* culture, following the same procedure just described.
6. Under Results, record your observations of the motility of the two bacterial organisms.
7. *Discard your slides in a container with disinfectant solution or in a biohazard container.*

Results

1. Make drawings in the following circles to show the *size, shape,* and *grouping* of each organism. Indicate with a check-mark below the circle whether it is *motile* or *nonmotile*. How does the size of each organism compare with that of the yeast cells in the preparation?

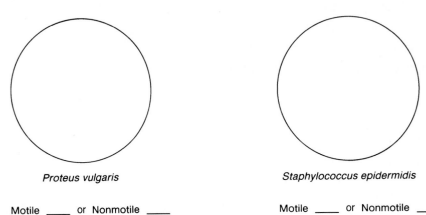

Proteus vulgaris

Motile _____ or Nonmotile _____

Staphylococcus epidermidis

Motile _____ or Nonmotile _____

2. In the following left-hand circle, draw the path of a single bacterium having true motility. In the right-hand circle, draw the path of a single nonmotile bacterium.

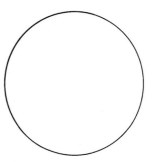

Path of a motile bacterium

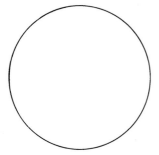

Path of a nonmotile bacterium

Basic Techniques of Microbiology

Questions

1. Define *acidic* and *basic dyes.* What is the purpose of each?

2. Why are specimens to be stained suspended in *sterile* saline or *distilled* water?

3. Which of the microscope objectives is most satisfactory for studying bacteria in stained preparations? In wet-mount preparations? Why?

4. List and define the basic shapes of bacteria. What are the dimensions of an average bacillus in micrometers? In centimeters?

5. List at least three types of bacteria whose names reflect their shapes and arrangements, and state the meaning of each name.

6. For what reason do we need to stain bacteria?

7. Examine **colorplates 1–8** and describe the morphology of the bacteria in each one.

8. How does true motility differ from Brownian movement?

9. What morphological structure is responsible for bacterial motility?

10. Why are wet-mount preparations discarded in disinfectant solution or a biohazard container?

11. How does a stained preparation compare with a wet mount for studying the morphology and motility of bacteria?

12. Can you save a wet mount to look at during a later class period? Why?

Gram Stain

Learning Objectives

After completing this exercise, students should be able to:

1. Explain two important differences between the cell walls of gram-positive and gram-negative bacteria.
2. Recognize why the Gram stain is important in clinical diagnosis.
3. Identify which step of the Gram-stain procedure could be left out and the reason why.
4. Explain why antibiotics such as penicillin that work by preventing the crosslinking of glycan molecules in peptidoglycan are more effective against gram-positive than gram-negative bacteria.
5. State the advantage of the Gram stain over a simple stain.

The simple staining procedure performed in Exercise 3 makes it possible to see bacteria clearly, but it does not distinguish between organisms of similar morphology.

In 1884, a Danish pathologist, Christian Gram, discovered a method of staining bacteria with pararosaniline dyes. Using two dyes in sequence, each of a different color, he found that bacteria fall into two groups. The first group retains the color of the primary dye: crystal violet (these are called *gram positive*). The second group loses the first dye when washed in a decolorizing solution but then takes on the color of the second dye, a *counterstain,* such as safranin or carbolfuchsin (these are called *gram negative*). An iodine solution is used as a *mordant* (a chemical that fixes a dye in or on a substance by combining with the dye to form an insoluble compound) for the first (crystal violet) stain.

Differences in the Gram-stain reactions are related to differences in the biochemical composition of bacterial cell walls. Gram-positive cell walls are composed of thick, tightly linked peptidoglycans (protein-sugar complexes) in which the crystal violet–iodine complex becomes trapped, thereby enabling the cells to resist decolorization. In contrast, gram-negative bacterial walls have a high concentration of lipids (fats) in their outer membranes and a thinner layer of peptidoglycan (fig. 4.1). The lipids dissolve in the decolorizer (alcohol, acetone, or a mixture of these) and are washed away along with the crystal violet–iodine complex. After decolorization, the now colorless gram-negative organisms take up the red counterstain and appear pink. The gram-positive organisms, which have retained the crystal violet–iodine complex, appear purple after decolorization and counterstaining. See **colorplates 1–6** and **colorplate 8** for examples of gram-positive and gram-negative bacteria.

The Gram stain is one of the most useful tools in the microbiology laboratory and is used universally. In the diagnostic laboratory, it is used not only to study microorganisms in cultures, but also to stain and examine smears made directly from clinical specimens. Direct, Gram-stained smears are read promptly to determine the relative numbers and morphology of bacteria in the specimen. This information is valuable to the physician in planning the patient's treatment before culture results are available. It is also valuable to microbiologists, who can plan their culture procedures based on their knowledge of the bacterial forms they have seen in the specimen.

The numerous modifications of Gram's original method are based on the concentration of the dyes, length of staining time for each dye, and composition of the decolorizer. Hucker's modification, to be followed in this exercise, is commonly used today. The choice of decolorizing

Figure 4.1 Structure of gram-positive and gram-negative cell walls. The gram-positive cell wall contains a thick layer of peptidoglycan, whereas the gram-negative cell wall has an outer membrane containing lipids. These unique characteristics are most likely responsible for the particular Gram-stain reaction of each type. The periplasmic space may not be visible in gram-positive bacteria.

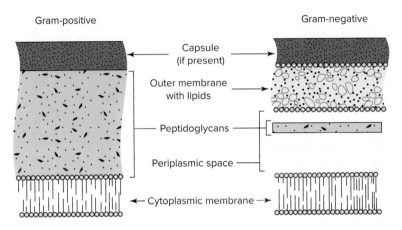

agent depends on the speed wanted to accomplish this step. The slowest agent, 95% ethyl alcohol, is used in this exercise to permit the student to gain experience with decolorization. Acetone is the fastest decolorizer, while an equal mixture of 95% ethyl alcohol and acetone acts with intermediate speed. The acetone-alcohol combination is probably the most popular in diagnostic laboratories.

An organism's appropriate Gram-stain reaction depends on the presence of an intact cell wall. As bacterial cultures age and the wall begins to disintegrate, gram-positive cells may not fully retain the crystal violet–iodine complex and will appear falsely gram-negative or as a mixture of gram-positive and gram-negative cells. In this latter case, they are referred to as gram variable. For this reason, cultures that have incubated longer than 24 or 48 hours or patient specimens that are not examined within a few hours of collection may give false readings when Gram-stained. Smears of bacterial colonies or organisms in patient specimens that are decolorized for too long (known as overdecolorization) may also give a false Gram-stain reading. In addition, some antimicrobial agents that act on the bacterial cell wall may change the morphology and Gram-stain reaction of the bacterial cells.

To properly judge the adequacy of a Gram-stained preparation, a smear of known gram-positive and gram-negative organisms is placed on the same slide as the unknown organism or patient specimen. This smear serves as the *control*, because on a well-stained smear, the appropriate Gram reaction (gram positive or gram negative) of these control organisms should be seen. If these organisms do not show their correct type of staining, the Gram reaction of the unknown is not reliable, and the staining procedure must be repeated.

Purpose	To learn the Gram-stain technique and to understand its value in the study of bacterial morphology
Materials	24-hour agar culture of:
	Staphylococcus epidermidis
	Enterococcus faecalis
	Neisseria sicca
	Saccharomyces cerevisiae (yeast)
	Bacillus subtilis
	Escherichia coli
	Proteus vulgaris
	Light suspension of *Staphylococcus epidermidis* and *Escherichia coli* for use as the control
	Specimen of simulated pus from a postoperative wound infection
	Wire inoculating loop
	Hucker's crystal violet
	Gram's iodine
	Ethyl alcohol, 95%
	Safranin
	Slides
	Marking pen or pencil and slide labels
	Test tube and staining racks
	Bibulous paper

Procedures

1. Label eight slides as follows: Se, Ef, Ns, Sc, Bs, Ec, Pv, and pus. Place a **small** drop of the "control" suspension on the left-hand side of each slide. On the right-hand side of all except the slide labeled "pus," prepare a **lightly turbid,** fixed smear of each bacterial culture. Follow the procedures outlined in Experiment 3.1, steps 1 through 6 (page 23). Refer also to the figure below. On the eighth slide, place a small drop of the simulated pus specimen, allow it to dry, and then fix the smear. If you are using unfrosted slides, make a code mark on the underside so that you can identify the slides after they are stained.
2. Stain each smear by the following procedures (this is Hucker's modification of the Gram stain). Refer also to table 4.1.
    a. Cover slide with crystal violet solution. Allow to stand for 1 minute (check with instructor; time varies with different batches of stain).
    b. Wash off with tap water.
    c. Cover slide with Gram's iodine (a mordant) solution. Leave for 1 minute.
    d. Wash off with tap water.
    e. Hold the slide at an angle over the staining rack and decolorize with ethyl alcohol (95%) dropwise until no more color washes off (usually 10–20 seconds). This is a most critical step. Be careful not to overdecolorize, as many gram-positive organisms may lose the violet stain easily and thus appear to be gram negative after they are counterstained.
    f. Wash off with tap water.
    g. Apply safranin (the counterstain) for 1 minute.
    h. Wash off with tap water.
    i. Drain and blot gently with bibulous paper. Air dry the slide thoroughly before you examine the preparation under the microscope.
3. When slides are dry, stick a label on each or use a permanent marker (like Sharpie) to label them as shown:

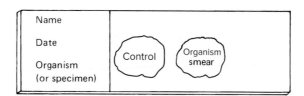

Table 4.1 Order of Reagents Used in the Gram-Stain Procedure

Reagent	Time*	Microscopic Appearance of Bacterial Cells After Reagent Application	
		Gram-Positive	Gram-Negative
1. Crystal violet	1 minute	Purple	Purple
2. Gram's iodine	1 minute	Purple	Purple
3. Decolorizer (alcohol, acetone, or a mixture of these two)	Until no more color washes out	Purple	Colorless
4. Safranin	1 minute	Purple	Pink

*Note that the timing will be specific for the reagents used by your instructor and that the slide is rinsed with tap water after each step.

4. Examine all slides under oil with the oil-immersion objective. Verify that each *control* smear contains gram-positive cocci and gram-negative rods.
5. Record observations of the Gram-stain reaction and bacterial morphology in the table under Results. Diagram with purple and pink pencils, if they are available.
6. Examine **colorplates 1–6** and **8,** noting which bacteria are gram positive or gram negative.

Results

Organism on Agar Plate

Name of Organism	Control Smear Shows Gram-Positive Cocci and Gram-Negative Rods (Yes or No)	Organism			
		Color (Purple or Pink)	Gram-Stain Reaction (Positive or Negative)	Cell Morphology (e.g., Cocci, Rods)	Diagram of Cell Morphology
Staphylococcus epidermidis					
Enterococcus faecalis					
Neisseria sicca					
Saccharomyces cerevisiae (yeast)					
Bacillus subtilis					
Escherichia coli					
Proteus vulgaris					

Pus Specimen

Describe Below Type(s) of Organisms Seen	Control Smear Shows Gram-Positive Cocci and Gram-Negative Rods (Yes or No)	Color (Purple or Pink)	Gram-Stain Reaction (Positive or Negative)	Cell Morphology (e.g., Cocci, Rods)	Diagram of Cell Morphology

Basic Techniques of Microbiology

Questions

1. What is the function of the iodine solution in the Gram stain? If it were omitted, how would staining results be affected?

2. What is the purpose of the alcohol solution in the Gram stain?

3. What counterstain is used? Why is it necessary? Could colors other than red be used?

4. If your control smear does not show gram-positive cocci and gram-negative rods, can you assume that the Gram-stain reaction of your "test" organism is correct? Why?

5. On the basis of Gram-stain reaction, can you distinguish species of:

    Staphylococcus and *Streptococcus?* _____

    Staphylococcus and *Neisseria?* _____

    Escherichia and *Proteus?* _____

    Escherichia and *Bacillus?* _____

6. What is the size of staphylococci in micrometers? In centimeters?

7. What is the advantage of the Gram stain over a simple stain such as methylene blue?

8. In what kind of clinical situation would a direct smear report from the laboratory be of urgent importance?

9. What is the mechanism of the Gram-stain reaction?

10. Describe at least two conditions in which an organism might stain gram variable.

Acid-Fast Stain

Learning Objectives

After completing this exercise, students should be able to:

1. Explain why the cell walls of acid-fast bacteria are unique.
2. Name two genera of bacteria that are acid-fast and two diseases caused by acid-fast bacteria.
3. Distinguish between the Kinyoun and Ziehl-Neelsen staining methods.
4. Explain why the acid-fast stain is considered to be a differential stain.
5. State what type of microscope must be used to view acid-fast bacteria stained with auramine dye.

Many organisms belonging to the genus *Mycobacterium* cause important human diseases. *Mycobacterium tuberculosis,* the agent of tuberculosis, continues to be a major public health problem worldwide. Tuberculosis is spread person to person by inhalation of infectious aerosols. It usually attacks the lungs but can affect any part of the body. Other important *Mycobacterium* species implicated in human disease are *M. kansasii, M. marinum,* members of the *M. avium* complex, and others. A large number of *Mycobacterium* species are found in the environment, and they are rarely pathogenic or have not yet been associated with infection. These are known as the saprophytic mycobacteria. Table 29.1 summarizes the mycobacteria that cause infectious diseases.

Members of the genus *Mycobacterium* contain large amounts of lipid (fatty) substances within their cell walls. These fatty waxes, also known as *mycolic acids,* resist staining by ordinary methods, so the diagnostic laboratory must use special stains to reveal them in clinical specimens or cultures (see also Exercise 29).

When these organisms are stained with a basic dye, such as carbolfuchsin, applied with heat or in a concentrated solution, the stain can penetrate the lipid cell wall and reach the cell cytoplasm. Once the cytoplasm is stained, it resists decolorization, even with harsh agents such as acid-alcohol, which cannot dissolve and penetrate beneath the mycobacterial lipid wall. Under these conditions of staining, the mycobacteria are said to be *acid fast* (see **colorplate 9**). Other bacteria whose cell walls do not contain high concentrations of lipid are readily decolorized by acid-alcohol after staining with carbolfuchsin and are said to be *non–acid fast.* One medically important genus, *Nocardia,* contains species that are *partially acid fast.* They resist decolorization with a weak (1%) sulfuric acid solution, but they lose the carbolfuchsin dye when treated with acid-alcohol. In the acid-fast technique, a counterstain is used to demonstrate whether or not the carbolfuchsin has been decolorized within cells and the second stain taken up, thus *differentiating* non-acid-fast from acid-fast organisms.

The original technique for applying carbolfuchsin with heat is called the *Ziehl-Neelsen stain,* named after the two bacteriologists who developed it in the late 1800s. The later modification of the technique employs more-concentrated carbolfuchsin reagent rather than heat to ensure stain penetration and is known as the *Kinyoun stain.* A fluorescence technique is now used more commonly in the mycobacteriology section of most clinical laboratories. In this method, the patient specimen is stained with the dye auramine, which fluoresces when it is exposed to an ultraviolet light source. Any acid-fast bacilli present on the slide take up this dye and fluoresce brightly against a dark background when viewed with a fluorescence microscope. Such smears are easier to read than those stained with a conventional carbolfuchsin-based stain. The smears can be examined at a lower power

(400×) than conventionally stained smears (1,000×) so that a larger area of the smear can be examined in a given time period. As with the interpretation of carbolfuchsin-stained smears, expertise is needed for interpreting fluorescent-stained smears. Not everything that fluoresces in such a preparation is necessarily a *Mycobacterium* species. Particulate matter in the specimen may retain the auramine stain. Especially when only a few organism-like structures are seen, it is important to pay careful attention to their morphology before interpreting them as mycobacteria (see **colorplate 9**).

Purpose	To learn the Kinyoun acid-fast technique and to understand its value when used to stain a clinical specimen
Materials	A young agar slant culture of *Mycobacterium phlei* or other saprophytic *Mycobacterium* sp.
	24-hour broth culture of *Bacillus subtilis*
	A sputum specimen simulating that of a 70-year-old man from a nursing home, admitted to the hospital with chest pain and bloody sputum
	Wire inoculating loop
	Gram-stain reagents
	Kinyoun's carbolfuchsin
	Acid-alcohol solution
	Methylene blue
	Slides
	Diamond glass-marking pencil
	Marking pencil or pen
	2 × 3-cm filter paper strips
	Slide rack
	Forceps
	Test tube and slide racks

Procedures

1. Prepare two fixed smears of each culture and two of the simulated sputum. In practice, the smears are fixed with methanol for 1 minute or are heat-fixed at 65° to 75°C to be certain any tuberculosis bacilli present are killed. To make smears of the agar slant culture, first place a drop of water on the slide, and then emulsify a **small** amount of the colonial growth in this drop.
2. Ring and code one slide of each pair with your marking pencil or pen, as usual.
3. The other slide of each pair must be ringed and coded with a diamond pencil. This device scratches the glass indelibly, so that the marks remain even during the prolonged staining process.
4. Gram-stain the set of slides marked with the pencil or pen in step 2.
5. Stain the diamond-scratched slides by the Kinyoun technique:
   a. Place the slides on a slide rack extended over a metal staining tray, if available.
   b. Cover smear with a 2 × 3-cm piece of filter paper to hold the stain on the slide and to filter out any undissolved dye crystals.
   c. Flood the slide with concentrated carbolfuchsin solution and allow to stand for 5 minutes (see fig. 5.1).
   d. Use forceps to remove filter paper strips from slides and place the strips in a discard container. Rinse slides with water and drain.
   e. Cover smears with acid-alcohol solution and allow them to stand for 2 minutes.
   f. Rinse again with water and drain.
   g. Flood smear with methylene blue and counterstain for 1 to 2 minutes.
   h. Rinse, drain, and air dry.
6. Examine all slides under oil immersion and record observations under Results. See **colorplate 9** for examples.

Basic Techniques of Microbiology

Figure 5.1 A smear flooded with carbolfuchsin is placed on a slide rack over a metal staining tray. ©*Verna Morton*

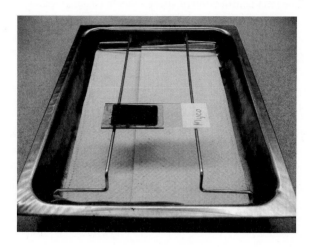

Results

Name of Organism	Visible in Gram Stain* (Yes, No)	Gram-Stain Reaction (If Visible)	Visible in Acid-Fast Stain (Yes, No)	Color in Acid-Fast Stain	Acid-Fast Reaction (If Visible)
Cultures					
Sputum specimen (describe organism)					

*Note: Some saprophytic mycobacteria may stain weakly gram positive or appear beaded in Gram-stained smears.

Special Stains

Capsules

In addition to the Gram and acid-fast stains, special stains are available to illustrate specialized microbial structures. These, however, are not used routinely in the clinical microbiology laboratory. For example, several microbial species possess an exterior capsule composed of carbohydrate or glycoprotein, which can be visualized microscopically by using a nonspecific *negative staining technique*. A drop of India ink or nigrosine is added to a suspension of the cells on a glass slide. These agents do not penetrate the cells or capsules but instead outline the capsules. The right-hand side of **colorplate 48** is a negative (India ink) stain of the encapsulated yeast *Cryptococcus neoformans*. Capsules contribute to microbial virulence by preventing phagocytic cells from ingesting and killing the organism.

Flagella

Bacterial flagella are tiny, hairlike organelles of locomotion. Originating in the cytoplasm beneath the cell wall, they extend beyond the cell, usually equaling or exceeding it in length. Their fine protein structure requires special staining techniques for demonstrating them with the light microscope. Because not all bacteria possess flagella, their presence, numbers, and pattern or arrangement on the cell may provide clues to identification of species. Most flagellar stains employ dyes and a mordant to "fix" the flagella and surface proteins of the cells. A precipitate then forms around these structures to make them visible when viewed under the microscope. In a similar way, a silver-plating technique is used to stain the very slender spirochetes (see **colorplate 7**).

Endospores

Among bacteria, endospore formation is most characteristic of two genera, *Bacillus* and *Clostridium*. The process of sporulation involves the condensation of vital cellular components within a thick, double-layered wall enclosing a round or ovoid inner body. The activities of the vegetative (actively growing) cell slow down, and it loses moisture as the endospore is formed. Gradually, the empty bacterial shell falls away. The remaining endospore is highly resistant to environmental influences, representing a resting, protective stage that is not easily stained by ordinary dyes. When stained with an endospore stain, the endospores take the color of the dye that is applied with heat (see **colorplate 8**).

Questions

1. What is a differential stain? Name two examples of such stains.

2. Is a Gram stain an adequate substitute for an acid-fast stain? Why?

3. When is it appropriate to ask the laboratory to perform an acid-fast stain?

4. In light of the clinical history (page 36) and your observations of the Gram- and acid-fast-stained smears, what is your tentative diagnosis of the patient's illness? How should this preliminary laboratory diagnosis be confirmed?

5. Are saprophytic mycobacteria acid fast?

6. Does the presence of acid-fast organisms in a clinical specimen always suggest serious clinical disease?

7. How should the acid-fast stain of a sputum specimen from a patient with suspected pulmonary *Nocardia* infection be performed?

8. Why is it important to know whether or not bacterial cells possess capsules, flagella, or endospores?

Culture Media

Learning Objectives

After completing this exercise, students should be able to:

1. Understand whether nutrient agar supports the growth of more or fewer organisms than a complex medium.
2. Describe where agar comes from and the special property of agar that makes it useful for inclusion in culture media.
3. State why Petri plates should be inverted during the incubation period.
4. Distinguish whether their agar plates have been contaminated before use.
5. Describe what would happen to plates poured with agar that is too hot or too cool, and state whether these plates could be used to cultivate bacteria.

Once the microscopic morphology and staining characteristics of a microorganism present in a clinical specimen are known, the microbiologist can make appropriate decisions as to how it should be cultivated and what biological properties must be demonstrated to identify it fully.

First, a suitable culture medium must be provided, and it must contain the nutrients essential for the growth of the microorganism to be studied (see Exercise 2). Most media designed for the initial growth and isolation of microorganisms are rich in protein components derived from animal meats. Many bacteria are unable to break down proteins to usable forms and must be provided with extracted or partially degraded protein materials (peptides, proteoses, peptones, amino acids). Meat extracts, or partially cooked meats, are the basic nutrients of many culture media. Some carbohydrate and mineral salts are usually added as well. Such basal media may then be supplemented, or *enriched,* with blood, serum, vitamins, other carbohydrates and mineral salts, or particular amino acids as needed.

For this exercise, the instructor has already prepared a basic nutrient broth medium and a nutrient agar from commercially available dehydrated stock mixtures containing all necessary ingredients. The term *nutrient broth* (or *agar*) refers specifically to basal media prepared from meat extracts, with a few other basic ingredients, but lacking special enrichment. The dehydrated powders were measured and added to a specified amount of distilled water. The mixtures were then boiled to completely dissolve the powder. The agar medium was then sterilized in the autoclave and maintained in a water bath at 50°C (this keeps the agar warm and melted, but not too hot to handle). We will learn how agar media are aseptically dispensed into Petri dishes and how tubes are filled with nutrient broth in preparation for sterilization.

Purpose	To learn how prepared culture media are dispensed for use in the microbiology laboratory
Materials	Prepared nutrient agar
	Prepared nutrient broth
	10-ml pipettes (cotton plugged)
	Test tubes (screw-capped or cotton plugged)
	Sterile Petri dishes
	Aspiration device for pipetting
	Test tube rack

Figure 6.1 Preparing a plate of agar medium by pouring melted sterile agar into it. *©Josephine A. Morello*

Procedures

1. The instructor will provide you with a flask of sterilized nutrient agar that has been maintained at 50°C. Remove the plug or cap with the little finger of your right hand and continue to hold it until you are sure you are done with the flask. Quickly pour the melted, sterile agar into a series of Petri dishes. The Petri dish tops are lifted with the left hand, and the bottoms are filled to about one-third capacity with melted agar (fig. 6.1). Replace each Petri dish top as the plate is poured. When the plates are cool (agar solidified), invert them to prevent condensing moisture from accumulating on the agar surfaces.
2. The instructor will provide you with a flask of 500 ml of nutrient broth and will demonstrate the use of the pipetting device. Using a pipette, dispense 5-ml aliquots of the broth into test tubes (plugged or capped). The instructor will collect the tubes and sterilize them.
3. Place inverted agar plates and tubes of sterilized nutrient broth (cooled after their return to you) in the 35°C incubator. They should be incubated for at least 24 hours to ensure they are sterile (free of contaminating bacteria) before you use them in Exercise 7.

Results

After at least 24 hours of incubation at 35°C, do your prepared plates and broths appear to be sterile?

_____

Record your observation of their physical appearance:

Plates: _____

_____

Broths: _____

Questions

1. Define a *culture medium*.

2. Which nutrient ingredients are required when preparing a culture medium for bacterial growth?

3. Which components of a culture medium are important for pH and buffering?

4. How are culture media sterilized before use and why are they sterilized?

5. Are nutrient broths and agars, as they have been prepared, suitable for supporting growth of all microorganisms pathogenic in humans? Explain your answer.

6. Discuss the relative value of broth and agar media in *isolating* bacteria from mixed cultures.

Streaking Technique to Obtain Pure Cultures

Learning Objectives

After completing this exercise, students should be able to:
1. Indicate why it is necessary to isolate individual colonies from a mixed bacterial growth.
2. Describe why it is important not to pass the inoculating loop back over the original inoculum when preparing a streaked plate.
3. State in which of the four quadrants of a streaked plate isolated colonies are expected to appear and why.
4. Recognize why a Petri dish should not be left open for an extended period.
5. Explain whether the Gram-stain reaction of bacteria can be predicted on the basis of their colony appearance on a streaked plate.

Large numbers of microorganisms that make up the normal, or indigenous, flora are found on the skin and many mucosal surfaces of the human body. When clinical specimens are collected from these surfaces and cultured on an agar medium, any pathogens and normal flora organisms that are present and supported by the culture conditions will grow as a *mixed culture*. Colonies of the pathogenic species must then be picked out of the mixture and grown as an isolated *pure culture*. Pure cultures are obtained by selecting a single colony from the mixture to *subculture* to another agar or broth medium. Because each isolated colony on a culture plate is a clone, derived by multiplication of a single bacterial cell, the subculture will contain only a single bacterial population. The microbiologist can then proceed to identify the isolated organism primarily by examining its colonial morphology characteristics and biochemical and immunological properties. Pure culture technique is critical to successful, accurate identification of microorganisms (see **colorplates 11–13**).

Purpose	A. To isolate pure cultures from a specimen containing mixed flora
	B. To culture and study the normal flora of the mouth
Materials	Nutrient agar plates*
	Blood agar plates
	Sterile swabs
	A mixed broth culture containing *Serratia marcescens* (pigmented), *Escherichia coli,* and *Staphylococcus epidermidis*
	A demonstration plate culture made from this broth, showing colonies isolated by good streaking technique
	Wire inoculating loop
	Glass slides
	Gram-stain reagents
	Test tube rack
	Marking pen or pencil

*If the plates you prepared in Exercise 6 are sterile and in good condition, they may be used in this experiment.

Procedures

A. Streaking a Mixed Broth Culture for Colony Isolation

1. With a marking pen or pencil, label the bottom of a nutrient agar plate with your name and the date. Be certain to write on only a small section of the plate (preferably near an edge) or you will not be able to examine the colonies that grow.
2. Make certain the contents of the broth culture tube are evenly mixed.
3. Sterilize the loop and let it cool in the air.
4. Place a loopful of broth culture on the surface of the labeled nutrient agar plate, near but not touching the edge. With the loop pressed *lightly* against the agar surface, streak the inoculum back and forth over approximately one-eighth the area of the plate; do not dig up the agar (fig. 7.1, area A). Sterilize the loop again and let it cool.
5. Rotate the open plate in your left hand so that you can streak a series of four lines back and forth, each passing through the inoculum and extending across one side of the plate (fig. 7.1, area B).
6. Sterilize the loop again and let it cool in air.
7. Rotate the plate and streak another series of four lines, each crossing the end of the last four streaks and extending across the adjacent side of the plate (fig. 7.1, area C).
8. Rotate the plate and repeat this parallel streaking once more (fig. 7.1, area D).
9. Finally, make a few streaks in the untouched center of the plate (fig. 7.1, area E). *Do not touch the original inoculum.*
10. Incubate the plate (inverted) at 35°C.

B. Taking a Culture from the Mouth

1. On a small area near one end of the bottom of a blood agar plate, write your name and the date with your marking pen or pencil.
2. Rotate a sterile swab over the surface of your tongue and gums.
3. Roll the swab over a small 1½-cm square of surface of a blood agar plate, near but not touching one edge (see fig. 7.1, area A). Rotate the swab fully in this area.
4. Discard the swab in a container of disinfectant or other appropriate biohazard container.
5. Using an inoculating loop, streak the plate as in figure 7.1.
6. Incubate the plate (inverted) at 35°C.

Figure 7.1 Diagram of plate streaking technique. The goal is to thin the numbers of bacteria growing in each successive area of the plate as it is rotated and streaked so that isolated colonies will appear in sections D and E.

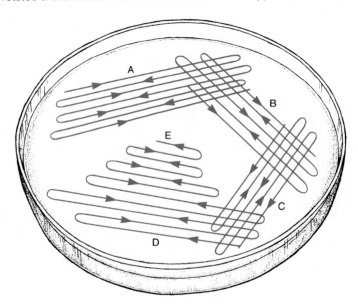

Basic Techniques of Microbiology

Results

A. Examination of Plate Streaked from Mixed Broth Culture

1. Examine the incubated nutrient agar plate carefully. Compare your streaking with that illustrated in figure 7.2a and b. Make a drawing showing the intensity of growth in each streaked area.

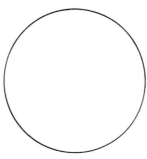

2. Describe each different type of colony you can distinguish (see fig. 2.1a, page 12).

3. Make a Gram stain of one isolated colony of each type present. Also prepare a Gram stain of the growth in the area where the initial inoculum was placed. (Note: when a stain is to be made of colonies growing on an agar medium, place a loopful of sterile water or saline on the slide first and then make a very light suspension of the picked growth in this drop. Allow to air dry, fix the slide by heat or methanol, and stain.)

Figure 7.2 Plate streaking. (a) Notice how the proper technique is designed to yield isolated colonies in areas D and E. (b) Poor streaking does not provide separation of colonies. ©Josephine A. Morello

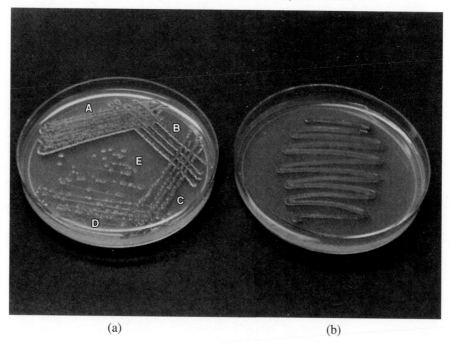

(a)   (b)

4. Record your observations in the table provided.

Single Colony	Colony Morphology	Pigment*	Gram Stain Reaction	Microscopic Morphology
Serratia marcescens				
Escherichia coli				
Staphylococcus epidermidis				
Area of initial inoculum				

Note: Keep the nutrient agar plate. You will work with it again in the next exercise.
*Pigmented strains of S. marcescens may not show pigment when incubated at 35°C.

B. Examination of Mouth Culture on Blood Agar Plate

1. How many different types of colonies can you find on the blood agar plate? _____
   Describe each.

2. Make a Gram stain of each of three different colonies. Record the Gram-stain reaction and morphology of each, and sketch its microscopic morphology in the circles.

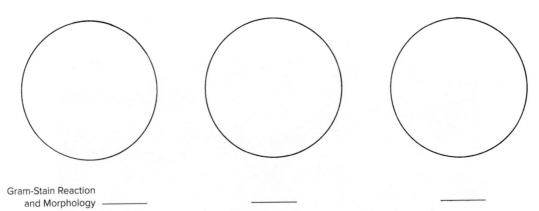

Gram-Stain Reaction
and Morphology ————   ————   ————

3. Discard the blood agar plate in a container marked CONTAMINATED or BIOHAZARD.

Questions

1. When an agar plate is inoculated, why is the loop sterilized after the initial inoculum is put on?

2. Distinguish between a *pure culture* and a *mixed culture.*

3. Define a *bacterial colony.* List four characteristics by which bacterial colonies may be distinguished.

4. Why does the streaking method you used to inoculate your plates result in isolated colonies?

5. Why was a blood agar, rather than a nutrient agar, plate used for the culture from your mouth?

6. Are the large numbers of microorganisms found in the mouth cause for concern? Explain.

Subculture Techniques

Learning Objectives

After completing this exercise, students should be able to:
1. Explain why it is necessary to make pure subcultures of organisms grown from clinical specimens.
2. Describe why pure subcultures are important in clinical diagnosis.
3. Determine whether a culture or subculture is pure.
4. List the kinds of specimens that may yield a mixed flora in bacterial cultures.
5. Describe why a Petri dish lid should not be placed face down on the bench top.

When primary isolation plates have been properly streaked by the method you learned in Exercise 7, individual colonies can be picked up on an inoculating loop or straight wire and inoculated to fresh agar or broth media. These new pure cultures of isolated organisms are called *subcultures.* If they are indeed pure and do not contain mixtures of different species, they can be identified in stepwise procedures, as you will see in later exercises. Pure subcultures are also required for antimicrobial susceptibility testing, as you will learn in Exercise 12.

EXPERIMENT 8.1   Subculture Technique (Picking Isolated Colonies for Pure Cultures)

Purpose	To obtain isolated colonies from streaked plate cultures and grow them as pure subcultures
Materials	Wire inoculating loop
	Nutrient agar plates (prepared in Exercise 6)
	Nutrient agar broth (prepared in Exercise 6)
	Nutrient agar plate cultures streaked in Exercise 7 containing isolated colonies of three bacterial species
	Marking pen or pencil
	Test tube rack

Procedures

1. Look again at figure 2.6 in Exercise 2. This figure illustrates the correct method of picking a single colony from the surface of a streaked plate.
2. Now open the nutrient agar plate you streaked in Exercise 7 from a mixed broth culture containing three organisms. Hold the exposed agar surface in good light so that you can see all surfaces and characteristics of individual isolated colonies.
3. With your sterilized, cooled loop held steady in your other hand, bring the loop edge down against the top surface of one isolated colony you have selected for pure subculture. Withdraw the charged loop (*don't touch it to anything!*) and close the streaked plate.
4. Inoculate a fresh, sterile nutrient broth by gently rubbing the charged loop against the inner wall of the tube, just beneath the fluid surface. When you bring the loop out of the tube, be sure it holds some of the broth.
5. Now use the loop to inoculate a fresh nutrient agar plate. Rub the inoculum onto a small area near the edge, sterilize the loop, and then go back and complete the streaking of the plate by using the technique illustrated in figure 7.1.
6. Inoculate two more agar plates, each with a different type of colony picked from your previous plate culture. Be certain all plates and tubes are labeled.
7. Incubate your new plate cultures (inverted) and broth cultures at 35°C.

Results

1. Examine your streaked nutrient agar plate subcultures and determine whether you have obtained pure cultures. In the following table, indicate the size, shape, and pigmentation of colonies on each plate. Make Gram stains of colonies on each subculture plate and record Gram-stain reactions and microscopic morphology in the table.

Organism	Colony Size (Diameter, mm)	Colony Shape	Pigment Color	Gram-Stain Reaction and Microscopic Morphology
Escherichia coli				
Serratia marcescens				
Staphylococcus epidermidis				

2. Examine your nutrient broth subcultures. Make Gram stains to determine whether they are pure. Describe your microscopic observations of each broth subculture.

_____

_____

_____

3. Are the organisms recovered in your plate and broth cultures the same as those you originally recorded in Exercise 7?

_____

If not, specify the differences: _____

_____

Questions

1. Why are plate cultures incubated in the inverted position?

2. How do you decide which colonies should be picked from a plate culture of mixed flora?

3. Is there a correlation between cell morphology and colony morphology? Explain.

4. When more than one colony type appears in a pure culture, what are the most likely sources of extraneous organisms?

PART TWO

Destruction of Microorganisms

Understanding how microorganisms can be destroyed is of utmost importance in patient-care situations, as well as in the laboratory. Many physical and chemical agents have antimicrobial activity; that is, they act against (*anti-*) microbes.

Physical Antimicrobial Agents

Among the numerous physical agents that are antimicrobial, heat is the most effective and reliable. It is also more efficient than most chemical antimicrobial agents because it can destroy microbes more rapidly under the right conditions.

When we use physical antimicrobial agents, we can be certain that *all* forms of microbial life are completely destroyed only when *sterilizing* techniques are used. The term *sterilization* is an absolute one; it means total, irreversible destruction of living cells. A number of *physical* environmental agents—such as ultraviolet or ionizing radiation, ultrasonic waves, or total dryness—exert stress on microorganisms and may kill them, but they cannot destroy large concentrations of microorganisms in a laboratory culture or a clinical specimen. Even small numbers of microorganisms may not be totally destroyed when exposed to ultraviolet rays or drying if, for example, they are protected by the fabrics contained in a clean surgical pack.

Ultraviolet light does not penetrate most substances, including fabrics, and therefore is used primarily to inactivate microorganisms located on surfaces. In microbiology laboratories, ultraviolet lamps are used inside of biological safety cabinets to decontaminate their surfaces, usually at the end of the day.

Filtration is another example of a physical method of sterilization that is used to remove bacteria, fungi, and their endospores from air or solutions. These microorganisms can be removed from air by passing the air through a high-efficiency particulate air (HEPA) filter. Solutions may be sterilized by passing the liquid through a special nitrocellulose filter that has a physical pore size of 0.45 to 0.2 microns. This pore size can remove all microorganisms except viruses from the solution. Most viruses are smaller in size and can pass through the filter. Filtration is used to sterilize solutions that contain biologic macromolecules, such as enzymes and other proteins, which are inactivated by high-temperature sterilization.

Heat is an excellent sterilizing agent when applied at high enough temperatures for an adequate period of time, because it effectively stops cellular activities. Depending on whether it is moist or dry, heat can coagulate cellular proteins (think of a boiled egg) or oxidize cell components (think of a burned finger or a flaming piece of paper). Heat is also nonselective in its effects on microorganisms (or other living cells), but we must bear in mind that this advantage is offset by its capacity to destroy all materials, whether living or not.

In Exercises 9 and 10, we shall see some examples of sterilization by use of moist and dry heat.

Chemical Antimicrobial Agents

When the application of heat is impractical, chemical agents are useful for destroying pathogenic microorganisms. A wide variety of chemical agents display antimicrobial activity to some degree. In considering their application to patient care, we may separate them into two general classes: (1) those that are useful for destroying pathogenic microorganisms in the environment (*disinfectants*) or on skin (*antiseptics*), and (2) those that may be administered to patients for treatment of infectious diseases (*antimicrobial agents,* often called *antibiotics*).

Many antimicrobial substances are too toxic to be used for patient therapy but are valuable as environmental disinfectants. These must be chosen carefully for the job to be done, because a given disinfectant usually does not kill all microbial pathogens. Each agent has a limited chemical mode of action, and microorganisms exposed to it may vary widely in their responses. Some microbes or their forms may succumb to its effects (such as vegetative bacterial cells) whereas others may not (such as bacterial endospores). In the experiments of Exercise 11, we shall study some of the many factors that influence the disinfection process.

The principles of heat sterilization and chemical disinfection are continuously applied to control infection and are an integral part of patient care. They are also essential to laboratory safety and must be understood before we proceed to the study of pathogenic microorganisms, with which diagnostic microbiology is concerned.

Antimicrobial agents, or *antibiotics*, are chemical substances that are naturally produced by a variety of microorganisms (primarily fungi and bacteria), or have been synthesized in the laboratory, or a combination of both. For example, scientists in pharmaceutical companies have made many chemical modifications of the penicillin molecule (a product of the fungus *Penicillium notatum*) to broaden its spectrum of activity. In strict use, *antibiotic* refers only to those antimicrobial substances produced by microorganisms, but the term is often used interchangeably with *antimicrobial agent*. Antimicrobial agents have inhibitory or lethal effects on many pathogenic organisms (especially bacteria) that cause infectious diseases. In purified form, they are administered to patients for their antimicrobial effects within the body. In general, each agent has special activity against one or more types of microorganisms (gram-positive bacteria, gram-negative bacteria, fungi, and some viruses).

Like disinfectants, antimicrobial agents have specific chemical modes of action, but the range of activity of antimicrobial agents is narrower. Therefore, as we shall learn in Exercise 12, the diagnostic microbiology laboratory tests the antimicrobial susceptibility of pathogenic bacteria so as to provide the physician with valuable information about the most clinically useful antimicrobial agent with which to treat a patient's infection specifically. At present, reliable tests for determining fungal and viral susceptibility to antimicrobial agents are under development and are only recently becoming generally available. In addition to the isolation and identification of pathogenic microorganisms that we shall study in sections of Part Three, antimicrobial susceptibility testing is one of the most important functions of the diagnostic microbiology laboratory.

Moist and Dry Heat

Learning Objectives

After completing this exercise, students should be able to:
1. Define thermal death point.
2. Define thermal death time.
3. Define sterilization.
4. Explain why heat may not always be a practical method of sterilization.
5. Describe the use of sterilization controls as used in this Exercise and their purpose.

In order to sterilize a given set of materials, the appropriate conditions of heat and moisture must be used with care given to applying the appropriate temperature for a sufficient time. Moist heat coagulates microbial proteins (including protein enzymes), inactivating them irreversibly. In the dry state, protein structures are more stable; therefore, the temperature of dry heat must be raised much higher and maintained longer than that of moist heat. For example, in a dry oven, 1 to 2 hours at 160° to 170°C is required for sterilization; however, with steam under pressure (the autoclave, see Exercise 10), only 15 minutes at 121°C may be needed. Note that these times and temperatures are just representative of the different cycles that may be used for sterilization. The choice of heat sterilization methods then depends on the heat sensitivity of materials to be sterilized.

EXPERIMENT 9.1   Moist Heat

It is possible to quantitate the response of microorganisms to heat by measuring the time required to kill them at different temperatures. The lowest temperature required to sterilize a standardized pure culture of bacteria within a given time (usually 10 minutes) can be called the *thermal death point* of that species. Conversely, the time required to sterilize the culture at a stated temperature can be established as the *thermal death time*.

Purpose	To demonstrate destruction of microorganisms by moist heat applied under controlled conditions of time and temperature
Materials	Tubed nutrient broths (4-ml aliquots)
	Nutrient agar plates
	Sterile 1.0-ml pipettes
	24-hour broth culture of *Staphylococcus epidermidis*
	6-day-old broth culture of *Bacillus subtilis*
	Wire inoculating loop
	Marking pen or pencil
	Test tube rack

Procedures

1. Set up a beaker water bath and heat to boiling.
2. Divide one nutrient agar plate in half by marking the bottom of the plate with a wax pencil or ink marker.
3. Label one-half of the plate *S. epidermidis* and the word *Control*. Streak a loopful of the *S. epidermidis* culture onto that half of the nutrient agar plate.

4. Repeat step 3 with the culture of *B. subtilis,* inoculating the second half of the plate.
5. Place the "control" plate in the 35°C incubator for 24 hours.
6. Divide two nutrient agar plates into four quadrants by marking the bottom of the plates with a wax pencil or ink marker. Label one plate *S. epidermidis* and the other *B. subtilis*. Label the four quadrants on each plate as follows: 5, 10, 15, and 20 minutes.
7. Take a pair of broth tubes and inoculate each, respectively, with 1 ml of *S. epidermidis* and *B. subtilis*. Place these tubes in the boiling water bath. Note the time.
8. Leave the pair of broth cultures in boiling water for 5 minutes. Remove the tubes carefully, being certain not to drip boiling water on your hand. Use a padded glove if available. Cool them quickly under running cold tap water. Streak a loopful of each boiled culture onto the quadrant of nutrient agar labeled 5 minutes.
9. Return the tubes to the boiling water bath for an additional 5 minutes. Begin timing when the water comes to a full boil. Cool the tubes as in step 8 then streak a loopful of each culture onto the quadrant of nutrient agar labeled 10 minutes.
10. Repeat step 9 twice more, streaking loopfuls of culture onto the quadrants of the plates labeled 15 and 20 minutes, respectively.
11. Incubate the subcultures from boiled tubes at 35°C for 24 hours.

Results

1. Read all plates for growth (+) or no growth (−). Record your results in the chart.

Culture	Minutes Boiled				Control
	5	10	15	20	
Staphylococcus epidermidis					
Bacillus subtilis					

2. State your interpretation of these results for each organism:

   Staphylococcus epidermidis:

   Bacillus subtilis:

EXPERIMENT **9.2**   **Dry Heat**

In this experiment, egg white (the protein albumin) is used to simulate microbial enzyme protein. The speed of the damaging reaction (coagulation) of moist and dry heat on protein will be observed.

Purpose	To compare the effects of moist and dry heat
Materials	Tubed distilled water (0.5-ml aliquots)
	Sterile 1.0-ml pipettes
	Clean tubes
	Dry-heat oven
	Egg white (albumin, a protein)
	Test tube rack

Procedures

1. Set up a beaker water bath and heat to boiling.
2. Set the dry-heat oven for 100°C.
3. Using a pipette, measure 0.5 ml of egg white into 0.5 ml of distilled water.
4. Place the tube into the boiling water bath and *immediately* begin timing. Observe until the egg white has coagulated, then record the elapsed time.
5. Using a pipette, measure 1.0 ml of egg white into a clean tube.
6. Place the tube into the dry-heat oven and *immediately* begin timing. Observe until the egg white has coagulated, then record the elapsed time.

Results

1. Elapsed time for protein coagulation in moist heat (boiling): _____

   Elapsed time for protein coagulation in dry heat (baking): _____

2. State your interpretation of the effect of moisture on protein denaturation: _____

   _____

EXPERIMENT **9.3**   **Incineration**

Purpose	To learn the effect of flaming with dry heat
Materials	Nutrient agar plates 24-hour broth culture of *Staphylococcus epidermidis* 6-day-old culture of *Bacillus subtilis* Wire inoculating loop Marking pen or pencil

Procedures

1. With your marking pen or pencil, section an agar plate into two parts.
2. Streak the *S. epidermidis* culture on one-half of the plate. Label this section *Control.*
3. Sterilize the loop, take another loopful of *S. epidermidis* culture and sterilize the loop again. When the loop is cool, use it to streak the second half of the plate. Label this section *Heated* and the name of the organism.
4. Repeat procedures 1 to 3 with the *B. subtilis* culture.
5. Incubate the plates at 35°C for 24 hours.

Results

1. Read for growth (+) or no growth (−) and record.

Organism	Control	Incineration
Staphylococcus epidermidis		
Bacillus subtilis		

2. State your interpretation of the effect of incineration on:

A bacterial vegetative cell (*Staphylococcus epidermidis*)

_____

_____

An endospore-forming organism (*Bacillus subtilis*)

_____

_____

Questions

1. How are microorganisms destroyed by moist heat? By dry heat?

2. Are some microorganisms more resistant to heat than others? If so, why?

3. Is moist heat more effective than dry heat? If so, why?

4. Why does dry heat require higher temperatures for longer time periods to sterilize than does moist heat?

5. What is the relationship of time to temperature in heat sterilization? Explain.

6. Would you recommend boiling or baking to sterilize a soiled surgical instrument? Why?

7. What kinds of clean hospital materials would you sterilize by baking? Why?

8. Name some hospital materials that can be sterilized by flaming without harming them.

The Autoclave

Learning Objectives

After completing this exercise, students should be able to:
1. Give an example of an autoclave that you might find in a home kitchen.
2. List the two most important factors in heat sterilization.
3. Explain how an endospore sample is used as a valid sterilization control.
4. Describe the factors that determine the time necessary to achieve steam-pressure sterilization and to achieve dry-heat oven sterilization.
5. Specify how pressure, time, and temperature are used in routine autoclaving.

The autoclave is a steam-pressure sterilizer. Steam is the vapor given off by water when it boils at 100°C. If steam is trapped and compressed, its temperature rises as the pressure on it increases. As pressure is exerted on a vapor or gas to keep it enclosed within a certain area, the energy of the gaseous molecules is concentrated and exerts equal pressure against the opposing force. The energy of pressurized gas generates heat as well as force. Thus, the temperature of steam produced at 100°C rises sharply above this level if the steam is trapped within a chamber that permits it to accumulate but not to escape. A kitchen pressure cooker illustrates this principle because it is, indeed, an "autoclave." When a pressure cooker containing a little water is placed over a hot burner, the water soon comes to a boil. If the lid of the cooker is then clamped down tightly while heating continues, steam continues to be generated but, having nowhere to go, creates pressure as its temperature climbs steeply. Unfortunately, this phenomenon has led to the development of pressure cookers for use as improvised explosive devices. This device may be used in the kitchen to speed cooking of food, because pressurized steam with its high temperature (120° to 125°C) penetrates raw meats and vegetables much more quickly than does boiling water or its escaping steam. In the process, any microorganisms that may also be present are similarly penetrated by the hot pressurized steam and destroyed.

Essentially, an autoclave is a large, heavy-walled chamber with a steam inlet and an air outlet (fig. 10.1). It can be sealed to force steam accumulation. Steam (being lighter but hotter than air) is admitted through an inlet pipe in the upper part of the rear wall. As it rushes in, it pushes the cool air in the chamber forward and down through an air discharge line in the floor of the chamber at its front. When all the cool air has been pushed down the line, it is followed by hot steam, the temperature of which triggers a thermostatic valve placed in the discharge pipe. The valve closes off the line and then, as steam continues to enter the sealed chamber, pressure and temperature begin to build up quickly. The barometric pressure of normal atmosphere is about 15 lb to the square inch. Within an autoclave, steam pressure can build to 15 to 30 lb per square inch *above* atmospheric pressure, bringing the temperature up with it to 121° to 123°C. Steam is wet and penetrative to begin with, even at 100°C (the boiling point of water). When raised to a high temperature and driven by pressure, it penetrates thick substances that would be only superficially bathed by steam at atmospheric pressure. In comparison to a dry-heat oven, which requires temperatures of 160° to 170°C, under autoclave conditions, pressurized steam kills bacterial endospores, vegetative bacilli, and other microbial forms quickly and effectively at temperatures much lower and less destructive to materials.

Temperature and *time* are the two essential factors in heat sterilization. In the autoclave (steam-pressure sterilizer), it is the intensity of *steam temperature* that sterilizes (pressure only

Figure 10.1 The autoclave. From Adrian N. C. Delaat, *Microbiology for the Allied Health Professionals,* 2nd ed. *Source: Adrian N. C. Delaat, Microbiology for the Allied Health Professionals, 2nd ed.*

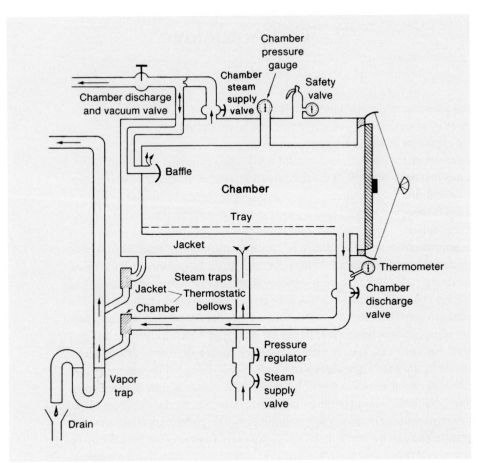

provides the means of creating this intensity), when it is given *time* measured according to the nature of the load in the chamber. In the dry-heat oven, the temperature of the hot air (which is not very penetrative) also sterilizes, but only after enough time has been allowed to heat the oven load and oxidize vital components of microorganisms without damaging materials. Table 10.1 illustrates the influence of pressure on the temperature of steam and, in turn, the influence of temperature on the time required to kill heat-resistant bacterial endospores. Compare these figures with those required for an average oven load—160°C for 2 hours, 170°C for 1 hour—and you will see

Table 10.1 Pressure-Temperature-Time Relationships in Steam-Pressure Sterilization

Steam Pressure, Pounds per Square Inch (above Atmospheric Pressure)	Temperature		Time (Minutes Required to Kill Exposed Heat-Resistant Endospores)
	Centigrade	Fahrenheit	
0	100°	212°	—
10	115.5°	240°	15–60
15	121.5°	251°	12–15
20	126.5°	260°	5–12
30	134°	273°	3–5

Destruction of Microorganisms

the efficiency of steam-pressure sterilization. Timing should not begin in either oven or autoclave sterilization until the interior chamber has reached sterilizing temperature.

The nature of the load in a heating sterilizing chamber greatly influences the time required to sterilize every item within the load. Steam *penetration* of thick, bulky, porous articles, such as operating room linen packs, takes much longer than does steam *condensation* on the surfaces of metal surgical instruments or laboratory glassware (quickly raised to sterilizing temperatures). The packaging of individual items (wrapped, plugged, or in a basket) also influences heat penetration, as does the arrangement of the total load in either an autoclave or an oven. In the autoclave, steam must be able to penetrate every surface of every item. In the oven, hot air must circulate freely around each piece in the load to bring it to sterilizing temperature. When sterilizing empty containers in a steam-pressure sterilizer, for example, it is important to consider that they contain cool air. Air is cooler and heavier than steam and cannot be permeated by it; therefore, microorganisms lingering within air pockets existing in or among items placed in an autoclave may survive steam exposure. For this reason, empty containers such as test tubes, syringes, beakers, and flasks should be laid on their sides so that the air they contain can run out and downward and be replaced by steam. Similarly, packaged materials should be positioned so that air pockets are not created among or between them.

Under routine circumstances, properly controlled steam pressure sterilization is accomplished in most hospitals under the following conditions of pressure, temperature, and time:

25 to 30 lb of steam pressure

133° to 135°C (271° to 275°F) steam temperature

5 to 25 minutes, depending on the nature of the load

Bacteriologic controls of proper autoclave function are essential to ensure that sterilization is being achieved with each run of the steam-pressure sterilizer. Preparations of heat-resistant bacterial endospores are commercially available for this purpose. Such preparations contain viable endospores dried on paper strips (fig. 10.2) or suspended in nutrient broth within a sealed ampule. When appropriately placed within an autoclave load, endospore controls can indicate whether the autoclave is operating efficiently and mechanically; individual item packaging is correct; and load arrangement permits sterilization of every item within the load.

The endospores of a bacterial species called *Bacillus stearothermophilus* provide a highly critical test of autoclave procedures because they are extremely resistant to the effects of moist or dry heat. As their name implies, they are heat (*thermo-*) -loving (*-philus*), but this also

---

Figure 10.2 Strips containing *B. stearothermophilus* endospores are placed in the autoclave with the material to be sterilized. After the autoclave cycle is completed, each strip is placed into a broth medium and incubated at 56°C. A second, control strip that has not been autoclaved is incubated in broth at the same time. The endospores on the control strip (left) have germinated, and the growing vegetative cells have made the broth turbid and changed the color of the pH indicator in the broth. The autoclaved endospores (right) have been successfully sterilized and, therefore, the broth remains clear and the original color. ©*Josephine A. Morello*

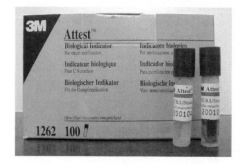

means that they require a higher incubation temperature than is optimal for most bacteria. The vegetative cells of *Bacillus stearothermophilus* grow best at *56°C* rather than at the 35°C temperature that is optimal for most pathogenic microorganisms. When dried on paper strips, these endospores provide a good test of oven sterilization techniques. When suspended in broth in sealed ampules, they are very useful for testing autoclave performance.

Bacillus stearothermophilus endospores on *paper strips* are packaged within paper envelopes that are placed within a load before heat sterilization. After sterilization, they are removed from their envelopes (aseptically), placed in appropriate nutrient broth, incubated at 56°C, and observed for evidence that they did or did not survive the sterilizing technique. *Sealed ampules* containing endospore suspensions are placed in an autoclave load (they cannot be used to test oven sterilization because they contain liquid), removed, and simply placed, without being opened, in an appropriate incubator (water bath or incubator at 56°C). Within a sealed ampule, endospores have been suspended in a nutrient broth also containing a pH-sensitive dye indicator. If endospores survive autoclaving and germinate again under incubation, vegetative bacilli begin to multiply in the broth. In the process, they use its nutrients, producing acid end products that cause the indicator to change color. They also produce turbidity in the medium.

When strips or ampules are used to test heat-sterilization technique, *unheated* strip or ampule controls must also be incubated to prove that the endospores were viable to begin with. At the completion of the incubation time, evidence of growth should be observed for the *control* but not the heated endospore preparations. If the heated test strips or ampules do not show growth by 24 to 48 hours, incubation should be continued for up to 7 days. The test then may be reported as negative, and the sterilization technique is assumed to have been effective. Patient-care materials included in the sterilized load are then safe to use. If, however, the endospores in the control preparation have not germinated, the test is considered unreliable, and the sterilized material cannot be assumed to be free of contaminating microorganisms. The sterilization procedure should be repeated with a new lot of strips or ampules.

Ampules containing liquid endospore suspensions must be kept refrigerated before use, because warm storage temperatures may permit endospore germination that could be wrongly interpreted. Dried endospore strips may be stored at room temperature because dry endospores are not likely to germinate.

In this exercise, you will have an opportunity to see the sterilizing effects of an autoclave.

Purpose	To illustrate the use and control of an autoclave
Materials	Commercially prepared strips or ampules containing *Bacillus stearothermophilus* endospores*
	Nutrient broth (if strips are used)
	Forceps (if strips are used)
	1.0-ml sterile pipettes
	Bulb or other aspiration device for pipette
	56°C water bath or incubator
	Phenol red glucose broth tubes
	6-day-old broth culture of *Bacillus subtilis*
	Marking pen or pencil

*Some commercially available paper strips (Steris Corp., "Spordex Biological Indicators") contain two types of endospores in combination: those of *Bacillus stearothermophilus* and also *Bacillus subtilis*. The latter are less heat resistant than endospores of *Bacillus stearothermophilus* and do not require a high incubation temperature to germinate (35° to 37°C is satisfactory for incubation of *Bacillus subtilis*). These combination strips can therefore be used in either a gas sterilizer, an autoclave, or an oven. In a gas sterilizer, the relatively low temperature will destroy *Bacillus subtilis* endospores but not those of *Bacillus stearothermophilus*. Strips used for this purpose may then be incubated at 35°C to test for the survival of *Bacillus subtilis* (the thermophile will not grow), while strips placed in an autoclave or oven load are incubated at 56°C to test for growth of *Bacillus stearothermophilus* (the mesophile will not grow).

Destruction of Microorganisms

Procedures

1. The instructor will discuss and demonstrate the operation of the autoclave.
2. Inoculate a tube of phenol red glucose broth with 0.1 ml of the *B. subtilis* culture (*finger* the pipette or use a pipette bulb). Label it *Unheated* and place it in the incubator at 35°C for 24 hours.
3. Submit the culture of *B. subtilis* for autoclaving at 15 lb, 121°C, for 15 minutes. Afterward, inoculate a tube of phenol red glucose broth with 0.1 ml of the autoclaved culture. Label it *Autoclaved*. Incubate the glucose broth at 35°C for 24 hours.
4. The instructor will demonstrate the use of endospore controls. An unheated *B. stearothermophilus* endospore preparation will be placed in a 56°C water bath or incubator. Another will be placed in the autoclave with your subculture of *B. subtilis* and then incubated.
   a. If strips are used, the paper envelope of one will be torn open, and the strip will be removed with heat-sterilized forceps and placed in nutrient broth incubated at 56°C. Another will be placed in the autoclave (in its envelope) and heated and then removed and placed in broth.
   b. If ampules are used, one will be placed (unheated, unopened) in the 56°C water bath or incubator. Another will be autoclaved and then incubated according to the manufacturer's directions.
5. After at least 24 hours of incubation of all cultures, read and examine them for evidence of growth (+) or no growth (−).

Results

1. Record culture results in the table.

Test Organism	Autoclave			Incubation Temperature	Appearance of Incubated Controls or Glucose Broth Cultures		
	Time	Temp.	Pressure		Color	Turbidity	Growth (+ or −)
Bacillus stearothermophilus Unheated control	X	X	X				
Autoclaved control							
Bacillus subtilis Unheated culture	X	X	X				
Autoclaved culture							

2. State your interpretation of these results.

3. State the method used for timing the autoclave in your experiment.

Questions

1. Define the principles of sterilization with an autoclave and with a dry-heat oven.

2. What pressure, temperature, and time are used in routine autoclaving in most hospitals?

3. Why is it necessary to use bacteriologic controls to monitor heat-sterilization techniques?

4. When running an endospore control of autoclaving technique, why is one endospore preparation incubated without heating?

5. Would a culture of *Escherichia coli* make a good bacteriologic control of heat-sterilization techniques? Why?

6. What characteristics of *Bacillus stearothermophilus* make it valuable for use as a control organism for heat-sterilization techniques? Explain.

7. What factors determine the choice of a paper strip containing bacterial endospores or a sealed ampule containing an endospore suspension for testing heat-sterilization equipment?

8. Would you choose a dry-heat oven, an autoclave, or incineration to heat sterilize the following items? State why.

Soiled dressings from a surgical wound: _____

_____

_____

Surgical instruments: _____

_____

_____

Clean laboratory glassware: _____

_____

_____

Clean reusable syringes: _____

_____

_____

9. Why should the results of endospore control tests be known before heat-sterilized materials are used for patient care?

Destruction of Microorganisms

Disinfectants

Learning Objectives

After completing this exercise, students should be able to:
1. Define a disinfectant.
2. Describe three mechanisms by which disinfectants work.
3. List three factors that can influence the effectiveness of a disinfectant.
4. Identify microbes that are most susceptible to disinfectants and explain why.
5. Identify microbes that are most resistant to disinfectants and explain why.

Disinfection is defined as the destruction of *pathogenic* microorganisms (not necessarily *all* microbial forms). It is a process involving chemical interactions between a toxic antimicrobial substance and enzymes or other constituents of microbial cells necessary for maintaining microbial life. A disinfectant must kill pathogens while it is in contact with them so that they cannot grow again when it is removed. In this case it is said to be -*cidal* (lethal), and it is described, according to the type of organism it kills, as bactericidal, virucidal, sporicidal, or simply germicidal. If the antimicrobial substance merely inhibits the organisms while it is in contact with them, the microorganisms may be able to multiply again when the disinfectant is removed. In this case, the agent is said to have *static* activity (it *arrests* growth) and may be described as bacteriostatic, fungistatic, or virustatic, as the case may be. According to its definition, a chemical disinfectant should produce irreversible changes that are lethal to cells.

Microorganisms of different groups are not uniformly susceptible to chemical disinfection. Tubercle bacilli (*Mycobacterium tuberculosis,* the etiologic agent of tuberculosis) are more resistant than most other vegetative bacteria because of their waxy cell walls, but of all microbial forms, bacterial endospores display the greatest resistance to both chemical and physical disinfecting agents. Fungal conidia (spores) are also somewhat resistant, although yeasts and hyphae (nonsporing fungal structures), like bacteria, succumb quickly to active disinfectants. Many bactericidal disinfectants also kill viruses, but the viral agents of hepatitis are very resistant.

Because microorganisms differ in their response to chemical antimicrobial agents, the choice of disinfectant for a particular purpose is guided in part by the type of microbe present in the contaminated material. Disinfectants that effectively kill vegetative bacteria may not destroy bacterial endospores, fungal conidia, mycobacteria, or some viruses. Other practical factors to consider when choosing a disinfectant include the exposure time and concentration required to kill microorganisms, the temperature and pH for its optimal activity, the concentration of microorganisms present, and the toxicity of the agent to skin or its effect on materials to be disinfected.

Table 11.1 summarizes the properties of some common disinfectants. Note that the concentrations at which they are effective vary among the agents and even for the same agent. Their activity against groups of microorganisms also differs, with most gram-positive and gram-negative vegetative bacteria being uniformly susceptible to disinfectants and bacterial endospores being the most resistant. Other important properties include whether they are inactivated by organic matter and soap and can maintain residual activity on various surfaces.

Some disinfectant use is limited to the environment, such as on inanimate surfaces (floors, furniture), whereas other compounds, such as 60 to 95% alcohols, 10% Betadine (a povidone-iodine compound), and 4% chlorhexidine gluconate (a modification of chlorhexidine) are effective for hand care use by health care workers. The chemical disinfection of skin is usually referred to as *antisepsis,* and the chemical compounds used for this purpose are called *antiseptics*. Proper attention to hand antisepsis is extremely important, because most hospital-acquired infections are transmitted to patients from the hands of health care workers.

In this exercise, we will study not only the effectiveness of disinfectants on cultures of vegetative (*Escherichia coli*) and endospore-forming (*Bacillus subtilis*) bacteria, but we will also determine the effectiveness of hand hygiene agents.

Purpose	To study the activity of some disinfectants and to learn the importance of time, germicidal concentration, and microbial species in disinfection
Materials	Nutrient agar plates
	Blood agar plates
	Sterile, empty tubes
	Sterile 5-ml pipettes (cotton plugged)
	Sterile 1.0-ml pipettes (cotton plugged)
	Bulb or other aspiration device for pipette
	5% sodium hypochlorite (bleach); 0.05% sodium hypochlorite
	Absolute ethyl alcohol, 70% ethyl alcohol
	3% hydrogen peroxide
	1% Lysol, 5% Lysol
	Iodophor (10% Betadine)
	Antiseptic mouthwash
	70% alcohol scrub
	4% chlorhexidine gluconate
	24-hour nutrient broth culture of *Escherichia coli*
	3- to 6-day-old broth culture of *Bacillus subtilis*
	Wire inoculating loop
	Test tube rack
	Marking pen or pencil

Destruction of Microorganisms

Table 11.1   Chemical Compounds Used for Disinfection and Their Selected Properties

Compound and Concentration	Selected Product	Effective Against*						Properties†		
		Bacterial Endospore	Vegetative Bacteria	M. tuberculosis	Fungi	Viruses		Effective in Organic Matter	Inactivated by Soap	Residual Activity
						Enveloped	Nonenveloped			
Alcohol 70%	Ethyl or Isopropyl	Poor	Good	Good	Good	Yes	No	Poor	No	Fair
Chlorhexidine 0.05–0.5%	Nolvasan	Poor	Good	Poor	Fair to Good	Limited	No	Fair	No	Good
Chlorine 0.01–5%	Clorox	Good	Good	Good	Good	Yes	Yes	Poor	No	Poor
Iodine/Iodophor	Tincture/10% Betadine	Poor	Good	Fair	Fair	Yes	Limited	Poor	Yes	Poor
Phenol 0.2–3%	pHisoHex	Poor	Good	Fair	Fair	Limited	No	Good	No	Poor
Quaternary ammonium 0.1–2%	Roccal-D	Poor	Good	Poor	Fair	Limited	No	Poor	Yes	Fair

*Effectiveness against microbial agents except viruses: good = effective; fair = moderate effect; poor = inferior effect. Inactivated by soap: yes = chemical compound has this property; no = chemical compound does not have this property. Residual activity: good = chemical compound has residual activity; fair = moderate residual activity; poor = inferior residual activity. Effectiveness against viruses: yes = effective; limited = moderate effect; no = not effective.

†Effectiveness in organic matter: good = effective; fair = moderate effect; poor = inferior effect.

Source: Modified from "Disinfectants and Their Properties," Appendix to *Morbidity and Mortality Weekly Report* 54, no. 4 (2005): 13.

Procedures

A. Activity of Disinfectants on Vegetative and Endospore-Forming Bacteria

1. Select one of the chemical agents provided. Label two sterile test tubes with the name and concentration of this disinfectant. Label one tube *E. coli* and the other *B. subtilis*. Pipette 3.0 ml of the selected solution into each tube.
2. With a marking pen or pencil, mark two nutrient agar plates on the bottom, dividing them into four sections each. Label both plates with the name of the disinfectant to be tested and its concentration. Label the four quadrants on each plate 2, 5, 10, and 15 min. Label one plate *E. coli* and the other *B. subtilis*.
3. To the 3 ml of disinfectant in the tube labeled *E. coli*, add 0.3 ml of the *E. coli* culture. Gently shake the tube to distribute the organisms uniformly. Note the time. Quickly perform step 4.
4. In the same way, add 0.3 ml of the *B. subtilis* culture to the appropriately labeled tube. Note the time.
5. At intervals of 2, 5, 10, and 15 minutes, transfer one loopful of the disinfectant-culture mixture to the section of the nutrient agar plates labeled with the corresponding time.
6. Label a nutrient agar plate with the word *Control* and divide the plate in half. Label one half *E. coli* and the other half *B. subtilis*. Inoculate the half labeled *E. coli* directly from the *E. coli* culture and the other half directly from the *B. subtilis* culture.
7. Incubate all tubes at 35°C for 48 hours.

B. Effectiveness of Hand Hygiene Practices

1. Choose one of the three hand hygiene products listed in the following chart: 70% alcohol scrub, 10% Betadine, or 4% chlorhexidine gluconate. Divide a blood agar plate in half. Label one half *Before* and the other half *After*.
2. Gently press the fingers of one hand onto the *Before* side of the plate.
3. Use the following chart to determine how you will sanitize your hands according to the product you have chosen.
4. Once your hands have been cleaned with the agent, press the fingers of the same hand you used on the *Before* side of the plate onto the *After* side of the blood agar plate.
5. Incubate all plates at 35°C for 24 hours.

Product*	Procedure	Time
70% Alcohol scrub	Add 3 to 4 ml of product to your hands and rub them sufficiently to cover all surfaces including palms, fingers, and fingernails. Do not rinse hands after the alcohol is dry but proceed to step 4.	Until all alcohol has evaporated
10% Betadine	Add 3 to 4 ml of product to your hands and rub in to cover all surfaces, as you do for alcohol. At the end of the treatment time, rinse thoroughly with tap water. Turn off the faucet with a paper towel and dry hands in air.	2 minutes
4% Chlorhexidine gluconate	Same as for Betadine	3 minutes

*A commercially available hand sanitizer, such as Purell (62% ethyl alcohol with inactive ingredients), may be substituted for one of the above products or added to this exercise.

Destruction of Microorganisms

Results

A. Examination of Plates to Determine Activity of Disinfectants on Vegetative and Endospore-Forming Bacteria

1. Observe the control plate and all other plate sections for growth (+) or absence of growth (−). Complete the table by recording your own and your neighbors' results with each disinfectant.

Disinfectant	Concentration	Organism	Time of Exposure (Minutes) 2	5	10	15	Control
Sodium hypochlorite	5%	E. coli					
		B. subtilis					
	0.05%	E. coli					
		B. subtilis					
Alcohol	Absolute	E. coli					
		B. subtilis					
	70%	E. coli					
		B. subtilis					
Hydrogen peroxide	3%	E. coli					
		B. subtilis					
Lysol	1%	E. coli					
		B. subtilis					
	5%	E. coli					
		B. subtilis					
Iodophor	10%	E. coli					
		B. subtilis					
Mouthwash	*	E. coli					
		B. subtilis					

*Check label of mouthwash bottle; fill in concentration of active ingredient.

2. State your interpretation of these results:

B. Examination of Plates Inoculated Before and After Hand Treatments

1. Examine the blood agar plate you inoculated before and after you treated your hands with one of the hand-cleaning agents. Estimate the number of colonies present and perform Gram stains on representative colonies. Record your results in the following chart and then record those of neighbors who used a different hand sanitizer.

Product	Before Treatment		After Treatment	
	Approximate Number of Bacteria	Approximate Type of Bacteria (Gram Stain and Morphology)	Approximate Number of Bacteria	Approximate Type of Bacteria (Gram Stain and Morphology)
70% Alcohol scrub				
10% Betadine				
4% Chlorhexidine gluconate				
*				

*Insert name of other, if used.

2. State your interpretation of these results.

CASE STUDY

A Lesson Learned—Follow Proper Procedures for the Disinfection of Equipment

In June of 2009, a congressional panel was pressing the Department of Veterans Affairs to determine whether more than 10,000 veterans may have been exposed to HIV and other infections at three southeastern hospitals. The reason was that proper procedures were not followed for disinfecting the endoscopic equipment used for colonoscopies and other gastrointestinal procedures. Those patients who may have been exposed needlessly to the contaminated equipment were asked to get blood tests for HIV and hepatitis. Six veterans who received the follow-up blood checks tested positive for HIV, 34 tested positive for hepatitis C, and another 13 tested positive for hepatitis B. Hospital administrators claim that all the mistakes were due to human error.

What do you think? Human error or human carelessness?

Destruction of Microorganisms

Questions

1. During their laboratory testing, if disinfectants are carried over into microbial cultures, could the results be affected? Explain.

2. Define *disinfection.* How does it differ from *antisepsis*?

3. Define bactericidal, bacteriostatic, virucidal, fungistatic.

4. Why are control cultures necessary in evaluating disinfectants?

5. What factors can influence the activity of a disinfectant?

6. Why do microorganisms differ in their response to disinfectants?

7. How can bacteriostatic and bactericidal disinfectants be distinguished?

8. What is an iodophor? What is its value?

9. Did you find the mouthwash you tested to be as effective as the other chemical agents included in this exercise? Explain any difference you observed.

10. Why are bacterial endospores a problem in the hospital environment?

11. Briefly discuss disinfection and antisepsis in relation to patient care.

12. In addition to using a hand scrub, what other precaution is necessary for health care personnel to prevent spreading infection by their hands when in the operating room?

13. Would you expect that long or artificial fingernails interfere with the effectiveness of hand scrubs? Why?

14. Why is it not suitable to wear rings in a hospital setting?

15. Why should hands be washed before and after dressing changes even though gloves must be worn during the procedure?

Antimicrobial Agent Susceptibility Testing and Resistance

Learning Objectives

After completing this exercise, students should be able to:
1. Define an *antimicrobial* agent.
2. Explain the meaning of antimicrobial resistance and susceptibility.
3. Specify why pure cultures are necessary for antimicrobial susceptibility testing.
4. Describe what would happen if a mixed culture were used in a susceptibility test.
5. State two mechanisms by which bacteria can acquire antibiotic resistance genes.

An important function of the diagnostic microbiology laboratory is to help the physician select effective antimicrobial agents for specific therapy of infectious diseases. When a clinically significant microorganism is isolated from the patient, it is usually necessary to determine how it responds in vitro to medically useful antimicrobial agents, so that the appropriate drug can be given to the patient. Antimicrobial susceptibility testing of the isolated pathogen indicates which drugs are most likely to inhibit or destroy it in vivo.

Susceptibility testing has shown that bacteria have become increasingly resistant to a wide variety of antimicrobial agents. Although new antibiotics continue to be developed by pharmaceutical manufacturers, the microbes seem to quickly find ways to avoid their effects. Two important bacteria that have developed resistance to multiple antimicrobial agents are *Staphylococcus aureus* strains, especially those resistant to the drug methicillin and its relatives, and *Enterococcus* spp. resistant to vancomycin. These organisms are referred to as methicillin-resistant *S. aureus* (MRSA) and vancomycin-resistant enterococci (VRE), respectively. Whereas MRSA strains were once thought to be confined to the hospital environment, they are increasingly seen as a cause of infections in nonhospitalized patients. In these cases, they are referred to as community-acquired MRSA and seem to differ in some important aspects from those MRSA strains that cause hospital-acquired infections. In particular, they are susceptible to more antibiotics than are hospital-acquired MRSA but are expected to become as resistant over time. They are responsible primarily for skin infections and may cause outbreaks among athletes whose sport involves close body contact, such as wrestlers.

Another disturbing trend is the increased resistance of hospital-acquired MRSA to vancomycin, one of the few agents now available to treat serious MRSA infections. Depending on the level of resistance, the organisms are referred to as vancomycin-intermediate *S. aureus* (VISA) or vancomycin-resistant *S. aureus* (VRSA). Methods for identifying staphylococci and enterococci are described in detail in Exercises 19 and 20, but antibiotic-resistant strains of both organisms play important roles in infections, especially those acquired by hospitalized patients. In like manner, gram-negative bacteria known as *Enterobacteriaceae*, which we will study in Exercise 23, are responsible for many common infections and have become resistant to a variety of antibiotics. These include agents in the carbapenem class of drugs, which are one of the few remaining options for treating infections caused by these bacteria. They are referred to as carbapenem-resistant *Enterobacteriaceae* or CRE. In 2016, a U.S. Senate report detailed a worldwide outbreak of CRE, related to duodenoscopes with design flaws that prevented them from being adequately cleaned and disinfected. At least 250 patients were affected. The laboratory must use methods to detect microbial resistance so that special precautions can be quickly instituted to prevent transfer of the resistant bacteria among patients and in the community. Some of the tests the laboratory has developed for this purpose are listed on page 88.

EXPERIMENT **12.1**   **Disk Agar Diffusion Method**

A testing method frequently used is the standardized *filter paper disk agar diffusion* method, also known as the CLSI (Clinical and Laboratory Standards Institute) *Bauer-Kirby* method. In this test, a number of small, sterile filter paper disks of uniform size (6 mm) that have each been impregnated with a defined concentration of an antimicrobial agent are placed on the surface of an agar plate previously inoculated with a standard amount of the organism to be tested. The plate is inoculated with uniform, close streaks to assure that the microbial growth will be confluent and evenly distributed across the entire plate surface. The agar medium must be appropriately enriched to support growth of the organism tested. Using a disk dispenser or sterile forceps, the disks are placed in even array on the plate, at well-spaced intervals from each other. When the disks are in firm contact with the agar, the antimicrobial agents diffuse into the surrounding medium and come in contact with the multiplying organisms. The plates are incubated at 35°C for 18 to 24 hours.

After incubation, the plates are examined for the presence of zones of inhibition of bacterial growth (clear rings) around the antimicrobial disks (see **colorplate 14**). If there is no inhibition, growth extends up to the rim of the disks on all sides and the organism is reported as resistant (R) to the antimicrobial agent in that disk. If a zone of inhibition surrounds the disk, the organism is not automatically considered susceptible (S) to the drug being tested. The diameter of the zone must first be measured (in millimeters) and compared for size with values listed in a standard chart (table 12.1). The size of the zone of inhibition depends on a number of factors, including the rate of diffusion of a given drug in the medium, the degree of susceptibility of the organism to the drug, the number of organisms inoculated on the plate, and their rate of growth. It is essential, therefore, that the test be performed in a fully standardized manner so that the values read from the chart provide an accurate interpretation of susceptibility or resistance. In some instances, the organism cannot be classified as either susceptible or resistant, but is interpreted as being of "intermediate" (I) susceptibility to a given drug. The clinical interpretation of this category is that the organisms tested may be inhibited by the antimicrobial agent provided that either (1) higher doses of the drug are

Table 12.1   Zone Diameter Interpretive Table

Antimicrobial Agent	Disk Concentration	Diameter of Inhibition Zone (mm)		
		R	I	S
Ampicillin[a]	10 µg	≤13	14–16	≥17
Cefoxitin	30 µg	≤14	15–17	≥18
Cephalothin	30 µg	≤14	15–17	≥18
Clindamycin[b]	2 µg	≤14	15–20	≥21
Ciprofloxacin	5 µg	≤15	16–20	≥21
Erythromycin[b]	15 µg	≤13	14–22	≥23
Gentamicin	10 µg	≤12	13–14	≥15
Penicillin G[b]	10 units	≤28	—	≥29
Penicillin G[c]	10 units	≤14	—	≥15
Piperacillin[d]	100 µg	≤14	15–20	≥21
Sulfonamides	250 or 300 µg	≤12	13–16	≥17
Tetracycline[b]	30 µg	≤14	15–18	≥19
Vancomycin[c,e]	30 µg	≤14	15–16	≥17

Source: Adapted from *Performance Standards for Antimicrobial Susceptibility Testing*—26th Informational Supplement (M100-S26), (CLSI 2016). The material is constantly being updated, and you can obtain the latest information from the Clinical and Laboratory Standards Institute.

Note: Zone sizes appropriate only when testing

[a]*Enterobacteriaceae*
[b]Staphylococci
[c]Enterococci
[d]*Pseudomonas aeruginosa*
[e]Staphylococcal isolates require vancomycin testing by the broth dilution method (see Experiment 12.2) instead of disk testing to determine whether they represent the emergence of a seriously resistant pathogen.

Destruction of Microorganisms

given to the patient, or (2) the infection is at a body site where the drug is concentrated; for example, the penicillins are excreted from the body by the kidneys and reach higher concentrations in the urinary tract than in the bloodstream or tissues. When an interpretation of I is obtained, the physician may wish to select an alternative antimicrobial agent to which the infecting micro-organism is fully susceptible, or additional tests may be necessary to assess the susceptibility of the organism more precisely.

Purpose	To learn the disk agar diffusion technique for antimicrobial susceptibility testing
Materials	Nutrient agar plates (Mueller-Hinton, if available)
	Tubes of sterile nutrient broth or saline (5 ml each)
	Antimicrobial disks (various drugs in standard concentrations)
	Antimicrobial disk dispenser (optional)
	McFarland No. 0.5 turbidity standard
	Sterile swabs
	Forceps
	24-hour plate cultures of *Staphylococcus epidermidis* and *Escherichia coli*
	Wire inoculating loop
	Marking pen or pencil
	Test tube rack

Procedures

1. Touch four to five colonies of *S. epidermidis* with your sterilized and cooled inoculating loop. Emulsify the colonies in 5 ml of sterile broth or saline until the turbidity is approximately equivalent to that of the McFarland No. 0.5 turbidity standard. Comparison with this standard ensures that the inoculum is neither too light nor too heavy.
2. Dip a swab into the bacterial suspension, express any excess fluid against the side of the tube, and inoculate the surface of an agar plate as follows: first, with back and forth motions, streak the *entire* surface of the plate with the swab (fig. 12.1, arrow direction 1). Then rotate the plate approximately 60° and streak the whole surface again in the direction shown in fig. 12.1 (arrow direction 2). Rotate the plate 60° once more and streak the entire plate (fig. 12.1, arrow direction 3). Finally, streak around the entire rim of agar (fig. 12.1, arrow direction 4). Discard the swab in disinfectant or a biohazard container.
3. Repeat steps 1 and 2 with the *E. coli* culture on a second nutrient agar plate.
4. Heat the forceps in the Bunsen burner flame or preferably, a bacterial incinerator, and allow to cool.
5. Pick up an antimicrobial disk with the forceps and place it on the agar surface of one of the inoculated plates. Press the disk gently into full contact with the agar, using the tips of the forceps.

Figure 12.1 Inoculating a disk agar diffusion susceptibility test plate. A swab moistened with a broth culture of the organism to be tested is used to streak the *entire* plate in three directions at approximately 60° angles to each other (arrows labeled 1, 2, and 3). The overlapping streaks are placed as close together as possible. Once the entire plate is streaked, the swab is rubbed around the entire rim of the plate, as shown by arrow 4.

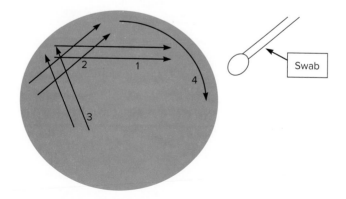

6. Heat the forceps again and cool.
7. Repeat steps 5 and 6 until about eight different disks are in place on one plate, spaced evenly away from each other. (If an antimicrobial disk dispenser is available, all disks may be dispensed on the agar surface simultaneously. Be certain to press them into contact with the agar using the forceps tips.)
8. Place a duplicate of each disk on the other inoculated plate, using the same procedures.
9. Invert the plates and incubate them at 35°C for 18 to 24 hours.

Results

Observe for the presence or absence of growth around each antimicrobial disk on each plate culture. Using a ruler with millimeter markings, measure the diameters of any zones of inhibition (see fig. 12.2) and record them in the chart. If the organism grows right up to the edge of a disk, record a zone diameter of 6 mm (the diameter of the disk). Compare the diameters of the inhibition zones with those in table 12.1 and record your interpretations of S, I, or R.

Antimicrobial Agent	Disk Concentration	Zone Diameter		E. coli (S, I, or R)	S. epidermidis (S, I, or R)
		E. coli	S. epidermidis		

Figure 12.2 Measuring the diameters of the inhibition zones around the antimicrobial disks on a disk agar diffusion plate. The zones are measured in millimeters and the diameters are compared with those in a CLSI chart to determine whether the organism is susceptible (S), of intermediate susceptibility (I), or resistant (R) to each antimicrobial agent. If no inhibition zone is present, as in the top right disk in the figure, the zone is read as 6 mm, the diameter of the disk.

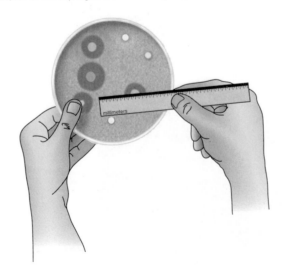

Destruction of Microorganisms

EXPERIMENT 12.2   Broth Dilution Method: Determining Minimum Inhibitory Concentration (MIC)

In certain instances of life-threatening infections such as bacterial endocarditis, or infections caused by highly or multiply resistant organisms, the physician may require a *quantitative* assessment of microorganism susceptibility rather than the qualita-
tive report of S, I, or R. The laboratory then tests the susceptibility of the organism to varying concentrations of one or more
appropriate antimicrobial agents. Twofold dilutions of each antimicrobial agent are prepared over a range of concentrations that
are achievable in the patient's bloodstream or urine (depending on the infection site) when standard doses of the drug are administered. In some cases, rather than preparing a full series of twofold dilutions, the organism is tested in only two or three
antimicrobial concentrations. In this *breakpoint dilution method,* the concentrations tested are chosen carefully to discriminate
between susceptible and resistant organisms.

     The antimicrobial dilutions may be prepared in a broth medium, or each concentration of antimicrobial agent to be
tested can be incorporated into an agar medium. In this *agar dilution method* many organisms (up to 32) can be tested on a
single agar plate, although several plates, each containing a different antimicrobial concentration, are needed to perform the
assay. When only a single organism is tested, the *broth dilution method* is more rapid and economical to perform. Most laboratories now use commercially available microdilution plates. These consist of multiwelled plastic plates prefilled with various
dilutions of several antimicrobial agents in broth (see **colorplate 15**). Using a multipronged device, the microwells are inoculated simultaneously with a standardized suspension of the test organism. Thus, the susceptibility of an organism to many antimicrobial agents may be readily tested.

     Regardless of the dilution method chosen, the results are interpreted in the same manner. After 18 to 24 hours of
incubation at 35°C, the broths or microdilution plates are examined for inhibition of bacterial growth. For each antimicrobial
agent in a dilution test, the *lowest* concentration that inhibits growth is referred to as the minimum inhibitory concentration,
or MIC.

     An additional method for determining MIC, the Etest, combines the features of the broth dilution and agar diffusion
tests. In this method, Etest strips are prepared in such a way that one side of the strip contains a continuous antibiotic concentration gradient along its length and the other side is printed with a MIC reading scale in µg/ml. After an agar plate has been
inoculated with the organism to be tested, as in the CLSI method, one or more Etest strips are placed (antibiotic side down)
on the surface of the agar plate. After overnight incubation, if the microorganism is susceptible to the drug tested, a symmetri-
cal ellipse of growth inhibition is centered along the strip (fig. 12.3). The MIC is read at the point where the zone edge intersects the strip. As with the disk agar diffusion method, in order to obtain accurate information from all these tests, variables
such as inoculum size, phase of organism growth, broth or agar medium used, and antimicrobial agent storage conditions must
be rigidly controlled.

Purpose	To learn the broth dilution method for antimicrobial susceptibility testing
Materials	Sterile nutrient broth (Mueller-Hinton if available)
	Sterile tubes
	Broth containing 128 µg of ampicillin per ml
	Sterile 1- and 5-ml pipettes
	Bulb or other aspiration device for pipette
	Tubes of sterile saline (5.0 and 9.9 ml per tube)
	Overnight plate culture of *Escherichia coli*
	Marking pen or pencil
	Test tube rack

Figure 12.3 The E-test. Strips impregnated with a continuous antibiotic concentration gradient are placed onto an agar plate that has been inoculated with the test organism. After overnight incubation, an ellipse forms along the strip where the organism growth has been inhibited. The point where the bottom edge of the ellipse intersects the strip is the MIC. ©Sirirat/Shutterstock RF

Procedures

1. Place nine sterile tubes in a rack and label them as follows:

Tube No.:	1	2	3	4	5	6	7	8	9
Label:	64	32	16	8	4	2	1	Growth Control	Sterility Control

2. With a 5-ml pipette add 0.5 ml of sterile nutrient broth to *each* tube.
3. Add 0.5 ml of the ampicillin broth to the first tube (fig. 12.4a). Discard the pipette. The concentration of ampicillin in this tube is 64 µg per ml.
4. Take a fresh pipette, introduce it into the first tube (64 µg per ml), mix the contents thoroughly, and transfer 0.5 ml from this tube into the second tube (32 µg per ml). Discard the pipette.
5. With a fresh pipette, mix the contents of the second tube and transfer 0.5 ml to the third tube (16 µg per ml).
6. Continue the dilution process through tube number 7. The eighth and ninth tubes receive no antibiotic.
7. After the contents of the seventh tube are mixed, discard 0.5 ml of broth so that the final volume in all tubes is 0.5 ml.
8. From the plate culture of *E. coli*, prepare a suspension of the organism in 5 ml of saline equivalent to a McFarland 0.5 standard (see Experiment 12.1).
9. With a sterile 1-ml pipette, transfer 0.1 ml of the *E. coli* suspension into a tube containing 9.9 ml of saline. Discard the pipette in disinfectant or a biohazard container.
10. With a fresh pipette, mix the contents of the tube well. Add 0.1 ml of this organism suspension to the antibiotic-containing broth tubes 1 through 7 and to the number 8 growth control tube (fig. 12.4b).
11. Shake the rack gently to mix the tube contents and place the tubes in the 35°C incubator for 18 to 24 hours.

Destruction of Microorganisms

Figure 12.4 Broth dilution technique. (a) The ampicillin-containing broth is serially diluted in tubes that have been filled with 0.5 ml of a nutrient broth. The growth and sterility control tubes receive no antibiotic. (b) After the antimicrobial dilutions are completed, 0.1 ml of the appropriately diluted organism suspension, in this case *E. coli*, is added to all except the sterility control tube. The number on each tube is the final concentration of ampicillin in that tube (µg/ml).

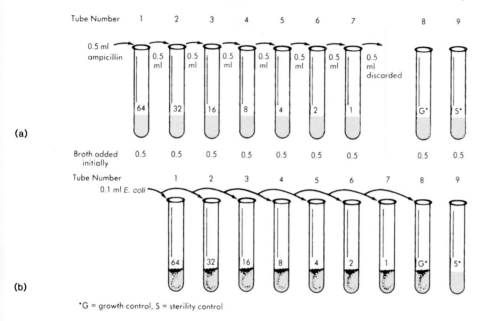

(a)

(b)

*G = growth control, S = sterility control

Results

1. Examine each tube for the presence or absence of turbidity. Record the results in the chart and indicate the MIC of ampicillin for the *E. coli* strain tested.

Antimicrobial Concentration (µg/ml)	64	32	16	8	4	2	1	Growth Control	Sterility Control
Growth (+ or −)									

MIC = _____ µg/ml

2. From the MIC, determine whether the *E. coli* strain is susceptible (S), of intermediate susceptibility (I), or resistant (R) to ampicillin, according to the following data from the CLSI Performance Standards for Antimicrobial Susceptibility Testing: S = ≤8, I = 16, R = ≥32.
The *E. coli* strain is _____.

EXPERIMENT 12.3   Bacterial Resistance to Antimicrobial Agents

The activity of antimicrobial agents is usually very specific, affecting primarily essential bacterial cell structures or biochemical processes. For example, penicillin interferes with bacterial cell wall synthesis, gentamicin inhibits protein synthesis, and sulfonamides block folic acid synthesis. During the years of widespread antimicrobial agent use, it became evident that bacteria have the ability to inactivate or in some way circumvent the activity of almost every known agent. Resistance to antimicrobial agents can result from a mutation in a gene on the bacterial chromosome, or by acquisition from another organism of a plasmid (extrachromosomal DNA) that bears one or more "resistance" genes (R-factor). Mutations occur at varying rates in the bacterial cell, usually between 1 in 10^4 to 1 in 10^{12} divisions. Bacteria are haploid; that is, they have only one unpaired chromosome in contrast to the multiple, paired (diploid) chromosomes seen in most higher organisms. Because of this haploid nature, recessive

Figure 12.5 Penicillin G (shown) and many of its derivatives are inactivated by a beta-lactamase (penicillinase). The enzyme breaks open the beta-lactam ring, which is a common part of the molecular structure of these antimicrobial agents.

Beta-lactam ring

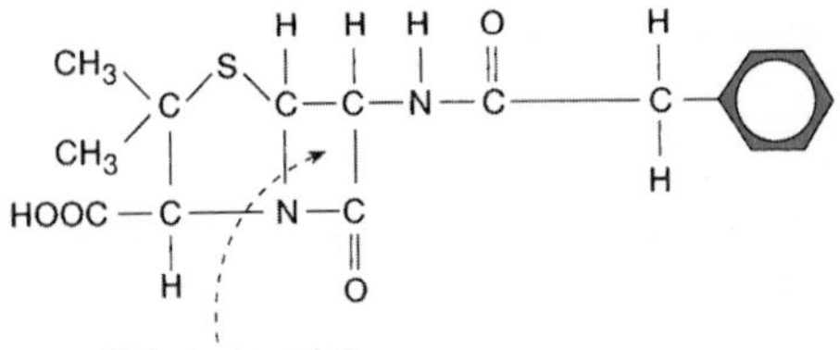

mutations do not occur as they do in diploid organisms and, therefore, bacterial mutations are more easily detectable. In addition, bacteria multiply to large populations rapidly (usually 10^9 cells in an overnight broth culture) so that mutants could be expected to arise in a short time. Most bacterial mutations go unrecognized, and the mutants may not survive unless the mutation provides a selective advantage for the cell, such as the ability to survive in the presence of an antimicrobial agent that is lethal for the wild-type (nonmutated) population.

In the case of R-factors, acquisition of this extrachromosomal DNA can suddenly render a previously susceptible bacterium resistant to multiple antimicrobial agents. One of the most common mechanisms of bacterial resistance is the production of specific enzymes (see also Exercise 15) that destroy antimicrobial agents before they can affect the bacterium. For example, penicillinase is an enzyme that inactivates penicillin by breaking open a particular structure on the penicillin molecule called a beta-lactam ring (penicillinase is also referred to as a beta-lactamase) (fig. 12.5). A gene on a plasmid in the bacterial cell provides instructions for formation of this enzyme. Although carriage of the penicillinase plasmid once appeared to be confined to certain strains of staphylococci and gram-negative bacilli, it is now found in some strains of bacteria that previously were considered to be universally susceptible to penicillin or its derivatives. These include *Haemophilus influenzae,* a cause of severe infections in children, before an effective vaccine became available, and *Neisseria gonorrhoeae,* the agent of gonorrhea.

Bacterial enzymes can also be responsible for resistance to antimicrobial agents other than penicillin. Gentamicin and chloramphenicol, for example, may be inactivated by enzymes specific for these drugs, but there are additional mechanisms by which bacteria may resist the action of certain antimicrobial agents. These include alterations in critical bacterial enzymes or proteins such that they can no longer be directly affected by the drug; or changes in the bacterial cell wall or membrane that make the cell less permeable, preventing entrance of the agent.

Routinely, the clinical microbiology laboratory tests for bacterial susceptibility or resistance by the methods described in Experiments 12.1 and 12.2. Alternatively, however, if you are interested only in the response of a given organism to a particular antimicrobial agent (e.g., *Neisseria gonorrhoeae* to penicillin), you can test the organism for its ability to produce a sufficient amount of an enzyme that specifically inactivates that drug. If the organism can be shown to possess the enzyme, it is considered to be resistant to the antimicrobial agent in question. One such test is illustrated in this experiment, using a penicillin-susceptible organism and one that is resistant to penicillin because it produces penicillinase. The test uses a filter paper disk containing the chromogenic (color-producing) cephalosporin, nitrocefin. Like penicillin, the cephalosporins are degraded by beta-lactamases. When the test disk is inoculated with a penicillinase-producing organism, the yellow nitrocefin is broken down to a red end product.

The laboratory is constantly challenged to incorporate new tests for detecting various resistance mechanisms. Examples are:

Detection of	Test
Carbapenemases	Modified Hodge test
Inducible clindamycin resistance	D-zone test
Extended-spectrum beta-lactamases (ESBL)	Phenotypic confirmatory ESBL test
Aminoglycoside synergy	High-level aminoglycoside screen test

Destruction of Microorganisms

Purpose	To detect penicillinase production by a test bacterial strain
Materials	Filter paper disks impregnated with nitrocefin for performing the beta-lactamase test
	Sterile water or saline
	Clean glass slides
	Forceps
	Plate culture of a penicillin-resistant *Staphylococcus aureus*
	Plate culture of a penicillin-susceptible *Bacillus subtilis*
	Wire inoculating loop

Procedures

1. Place two small drops of water or saline on the surface of a clean glass slide.
2. Pick up a beta-lactamase test disk with your forceps and place it in contact with one drop of fluid.
3. Repeat this procedure with a second disk, placing it on the second drop of fluid. Do not oversaturate the disks.
4. With your sterilized and cooled inoculating loop, pick up a portion of a *B. subtilis* colony and rub it across the surface of the first disk.
5. Rub a portion of a *S. aureus* colony across the surface of the second disk.
6. Observe the areas on the beta-lactamase test disks where the organisms were inoculated for up to 30 minutes. A positive result is usually seen within 3 to 4 minutes.

Results

1. A change in the color of the bacterial growth rubbed on the disk from yellow to red is a positive test indicating degradation of nitrocefin.
2. Record your results in the chart.

Organism	Color on Strip After 30 Minutes	Penicillinase + or −
Bacillus subtilis		
Staphylococcus aureus		

Questions

1. Describe the Bauer-Kirby test.

2. Explain the significance of colonies observed growing in the zone of inhibition in a Bauer-Kirby test.

3. What is the broth method for determining minimum inhibitory concentration (MIC)?

4. List three factors that can influence the accuracy of the test.

5. If a McFarland 0.5 standard contains 1×10^8 organisms per milliliter, how many bacteria were added to each ampicillin-containing tube in Experiment 12.2?

6. When performing a broth dilution test, why is it necessary to include a growth control tube? A sterility control tube?

7. How can the *minimum bactericidal concentration* of an antimicrobial agent be determined from an MIC assay?

8. Could an organism that is susceptible to an antimicrobial agent in laboratory testing fail to respond to it when that drug is used to treat the patient? Explain.

9. Are antibacterial agents useful in viral infections? Explain.

10. Why is it better to use the word *susceptible* rather than the word *sensitive* in describing an organism's response to a drug? When speaking of the patient, what does the term *drug sensitivity* mean?

11. Describe a mechanism of bacterial resistance to antimicrobial agents, other than resistance gene acquisition.

12. If the laboratory isolates *S. aureus* from five patients on the same day, is it necessary to test the antimicrobial susceptibility of each isolate? Why?

13. What are three actions that could prevent antibiotic resistance?

PART THREE

Diagnostic Microbiology in Action

General Considerations

The laboratory diagnosis of infectious disease **begins with the collection of a clinical specimen for laboratory testing. This may involve smear examination, culture, or, in some cases, the use of antigen or molecular test methods.** The specimen must be the *right* one, collected at the *right* time, transported in the *right* way to the *right* laboratory. On the basis of experience and standard practices, the laboratory must then use the best techniques and methods for rapid isolation and identification or for other detection of any pathogenic microorganisms present in the specimen. In Parts Three and Four of this manual, we shall see how specimens from various sites of the body are cultured, the normal flora they may contain, and how the most common pathogens are recognized. An understanding of the laboratory work flow, the culture timing, and sequence of tests performed is important for personnel caring for patients with infectious diseases. You will come to appreciate how preliminary information, transmitted from the laboratory as the culture work-up proceeds, and the final culture report are valuable for patient management.

Learning the terminology of laboratory reports and microbial nomenclature is also important so that you can recognize their significance immediately. For example, you will learn to understand that a report of "alpha-hemolytic *Streptococcus*" in a throat culture simply represents normal flora and has little, if any, clinical importance. In contrast, a report of "beta-hemolytic *Streptococcus* group A" signifies that the patient has a strep throat and must be treated as soon as possible to prevent possible sequelae from developing. Certain other reports, such as methicillin-resistant *Staphylococcus aureus* in a wound infection or *Clostridium difficile* in a stool specimen, indicate that the patient has a serious, readily transmissible disease, and that precautions must be instituted to prevent the spread of the disease. The exercises in Part Three are designed to familiarize you with principles you can use in your professional work, rather than providing comprehensive details of clinical microbiology.

Microbiology at the Bedside

Collection and Transport of Clinical Specimens for Culture

The first step in obtaining an accurate laboratory diagnosis of an infectious disease is the proper collection of an appropriate clinical specimen. An inadequate specimen may lead to a misdiagnosis. Most clinical laboratories provide a handbook that describes the appropriate specimen collection and transport methods. You should consult this handbook, which may be in printed form or available on the laboratory's website, whenever you are not certain of the proper procedures to use for this critical task. You can also contact the laboratory directly for information.

The following general rules apply:

1. *Before* collecting the sample, label the container with the patient's name, the date, and source of the specimen (this may be a computer-generated label).
2. Use strict aseptic technique throughout the procedure.
3. Cleanse your hands before approaching the patient. Wear disposable gloves when collecting and handling the specimen. Once the specimen is collected, remove the gloves and discard them in a biohazard container. Cleanse your hands again.
4. Collect the specimen at the *optimum* time, as ordered by the physician. In some patients, such as those with pneumonia caused by *Streptococcus pneumoniae,* the organism is present in the bloodstream only early in the infection. In patients with typhoid fever, *Salmonella typhi* is found in blood, urine, or feces at different times during the course of infection. Failure to collect the specimen at the appropriate time may result in a *negative* culture.
5. Make certain the specimen is *representative* of the infectious process. If pneumonia is suspected *sputum,* not saliva, must be collected in addition to blood. Pus from an abscess or wound should be aspirated by needle and syringe from the depth of the lesion, not swabbed from the surface.
6. Collect or place the specimen aseptically in an appropriate sterile container provided by the laboratory *for this purpose only.* If the specimen is contaminated in the process, it may be useless.
7. Make certain the *outside* of the specimen container is clean and not contaminated before sending it to the laboratory. If the container has been soiled on the outside, wipe it carefully with an effective disinfectant so that it will not transmit infection to those who handle it further.
8. Make certain the container is tightly closed so that its contents do not leak out during transport to the laboratory.
9. Check whether enough material has been collected for the laboratory to perform all tests requested. Consult the laboratory handbook for this information.
10. Place the container in a paper bag for transport, or if it is to be sent through a pneumatic tube system, place it in a sealable plastic bag.
11. Attach a laboratory request slip to the outside of the bag. Be certain that the patient's name on the request slip matches that on the specimen container. Otherwise, the laboratory will not accept the specimen. The slip should be completely and legibly filled in with all necessary information, including the suspected clinical diagnosis and the name of and contact information for the physician who orders the test. In many institutions, the complete information is entered into a computer and specimen identification is made by placing a barcoded label with all required information onto the specimen container.
12. Arrange for immediate transport of the specimen to the appropriate laboratory. Keep in mind that many pathogenic microorganisms are delicate and do not adjust well to conditions in the environment. If the specimen is not *cultured* promptly and culture media incubated at once, the pathogen(s) may die. In the case of an unavoidable delay in delivery, specimens must be stored according to the laboratory's directions.
13. *Cleanse your hands.*

Precautions for Handling Specimens and Cultures

In 2007, the Centers for Disease Control and Prevention (CDC) updated the 1996 publication describing *standard precautions* to reduce the risk of transmitting microorganisms from recognized and unrecognized sources of infection in hospitals, nursing homes, and other health care facilities. The precautions assume that all patients may have a transmissible infectious agent and should be cared for as such. These precautions were proposed initially in 1985 to address concerns about transmission by blood of HIV (the human immunodeficiency virus), the agent of AIDS (acquired immunodeficiency syndrome). The precautions now apply not only to blood but to all body fluids, secretions, and excretions (except sweat) regardless of whether they contain visible blood and also to nonintact skin and mucous membranes. These precautions include always taking care to avoid needle-stick injuries and washing hands thoroughly if contaminated with blood, other body fluids, or potentially contaminated articles. Workers may also need to wear gloves, masks, and protective clothing to provide a barrier to the exposure source. See Exercise 34 for a further discussion of CDC precautions. In the clinical microbiology laboratory, patient specimens are always processed in a biological safety cabinet; gloves are always worn when handling specimens; and gowns, masks, and protective eyewear are available for other procedures in which there is a risk of transmission of HIV or other highly transmissible agents.

In many of the following exercises, you will see and handle cultures of microorganisms that might be found in clinical specimens. Review the "Safety Procedures and Precautions" on pages xi to xii, as well as the "General Laboratory Directions" on pages xii to xiii. Note that *any* microorganism is potentially pathogenic if it enters a susceptible host. All cultures must be treated with care and respect.

In the laboratory, the basis for laboratory safety is to understand the rules for appropriate behavior and to know how to carry out culture procedures. A laboratory in which microorganisms are understood and controlled may be safer than a hospital, home, or public place where infection exists but is unrecognized. These rules are restated for emphasis:

1. No eating, drinking, or smoking in the laboratory.
2. Cleanse your hands carefully before and after each session.
3. Use disinfectant to keep your bench-top clean.
4. Keep your bench area free of clutter.
5. Never put contaminated items (inoculating loops, pipettes) down on the bench.
6. Tie long hair back out of the way.
7. Conduct yourself quietly.
8. Never take cultures out of the laboratory.
9. If a culture is spilled or broken, do not panic. Cover the area with a paper towel and flood it with disinfectant. Call the instructor.
10. Be assured that microorganisms do not fly, swim, or bombard. They can be controlled by good technique.

Normal Flora of the Body

Under normal circumstances, the human skin and mucous membranes are sterile only before birth. During passage through the birth canal, superficial tissues come in contact with microorganisms, become colonized by them, and remain so throughout life. Ordinarily, this colonization is harmless and may even be beneficial. Except under unusual conditions, the colonizing organisms (referred to as commensals or normal or indigenous flora) do not invade living cells and may even perform useful functions for the human host. These functions include production of metabolic substances that are used by the host (e.g., vitamin K) and competition with potential pathogens that may enter their body habitat and try to establish themselves.

Only the superficial body tissues are colonized with normal flora. These include the top layers of skin and the mucous membranes that line the upper portions of the respiratory tract (mouth, pharynx), the lower portions of the female genital tract (vagina), the lower portions of the intestinal tract (small and large intestine), and the first portion of the urethra in males and females. The microbial normal flora change with age, human activity, and environmental influences. *Everything* that influences the host also influences the host's normal flora, including physiological development, food or medications ingested, climate, clothing, and living habits. The microbial flora of different areas vary according to conditions of growth available at those sites.

Infectious diseases are usually established when pathogenic microorganisms enter by invading skin or other body surface tissues (respiratory, intestinal, urogenital tracts). *Direct* entry into deep tissues results only from trauma of some kind, the bite of an insect, insertion of a contaminated needle or instrument, or other type of penetrating injury. In the exercises that follow, we shall study some of the normal flora of the body, as well as some of the pathogenic bacteria that cause infectious disease.

Principles of Diagnostic Microbiology

In Exercises 13 through 18, you will learn how to culture clinical specimens and identify isolated microorganisms by classic culture methods. You will also learn the principles of antigen detection, nucleic acid assays, and the newer MALDI-TOF method for microorganism detection and identification. Then you will become familiar with how these methods are used to study the microbiology of the respiratory tract (Exercises 19–22), intestinal tract (Exercises 23 and 24), urinary and genital tracts (Exercises 25 and 26), and blood (Exercise 27).

Many of the methods in common use today are the same as those developed more than a century ago during the time of the renowned early bacteriologists, Louis Pasteur (1822–1895) and Robert Koch (1843–1910). Since then, however, a great deal has been learned about the biochemical, immunologic, and molecular characteristics of microbes. This knowledge has greatly improved the speed, ease, and accuracy with which today's microbiologists identify pathogenic microorganisms.

Even with technical advances that allow rapid microbial detection and identification (sometimes directly in the patient specimen), an understanding of the metabolic behavior of microorganisms in culture is essential. In most instances, prompt, accurate recognitions of pathogenic species is still carried out by choosing appropriate culture media for primary isolation of these organisms from clinical specimens and by selecting appropriate tests to determine their characteristic metabolic behavior.

Primary Media for Isolation of Microorganisms

Learning Objectives

After completing this exercise, students should be able to:

1. Distinguish between selective and differential media.
2. Explain the importance of inhibiting the growth of normal flora microorganisms in a clinical specimen.
3. List the organisms that are able to grow on the following culture media: chocolate agar, mannitol salt agar, MacConkey agar, Hektoen enteric agar.
4. Name the chemical component in phenylethyl alcohol agar that inhibits the growth of gram-negative bacteria.
5. Identify a culture medium that can be used to select for the growth of gram-positive bacteria and one that selects for gram-negative bacteria.

As we have seen, many clinical specimens contain a mixed flora of microorganisms. When these specimens are set up for culture, if only one isolation plate was inoculated, a great deal of time would be spent in subculturing and sorting through the bacterial species that grow out. Instead, the microbiologist uses several types of primary media at once (i.e., a battery) to culture the specimen initially. In general, the primary battery has three basic purposes: (1) to culture all bacterial species present and see which, if any, predominates; (2) to differentiate species by certain growth characteristics or biochemical responses to ingredients of the culture medium; and (3) to selectively encourage growth of those species of interest while suppressing the growth of normal flora.

The basic medium on which a majority of bacteria present in a clinical specimen will grow contains agar enriched with blood and other nutrients required by pathogens. The blood, which provides excellent enrichment, is obtained from animal sources, most often from sheep. In the past, even human blood (usually obtained from outdated collections in blood banks) was used occasionally for preparing certain types of culture media. However, this practice has been abandoned and is now prohibited. The reason is the potential health risk of acquiring bloodborne pathogens, such as HIV and hepatitis B or C, by handling or working with potentially contaminated human blood.

In addition to basic nutrients, *differential media* contain one or more components, such as a particular carbohydrate, that can be used by some microorganisms but not by others. If the microorganism uses the component during the incubation period, a change occurs in an indicator that is also included in the medium (see **colorplates 16** and **17**), which results in a color change in the medium.

Selective media contain one or more components that suppress the growth of some microorganisms without seriously affecting the ability of others to grow. Such media may also contain ingredients for differentiating among the species that do survive.

When a battery of several culture media such as those just described is streaked upon receipt of a clinical specimen, the first results indicate what types of bacteria are present, in general how many, and which did or did not use the differential carbohydrate. Also, the species of particular interest on the selective medium (if that species was present in the specimen) has been singled out and differentiated. Thus, the process of identification of isolated pathogens is already well under way after 24 hours of incubation of specimen cultures.

Table 13.1 Culture Media for the Isolation of Pathogenic Bacteria from Clinical Specimens

Media	Classification	Selective and/or Differential Agent(s)	Type(s) of Organisms Isolated
Chocolate agar	Enriched	1% hemoglobin and supplements	Most fastidious pathogens such as *Neisseria* and *Haemophilus*
Blood agar plate (BAP)	Enriched and differential	5% defibrinated sheep blood	Almost all bacteria; differential for hemolytic organisms
Mannitol salt agar (MSA)	Selective and differential	7.5% NaCl and mannitol for isolation and identification of most *S. aureus* strains	Staphylococci and micrococci
MacConkey agar (MAC)	Selective and differential	Lactose, bile salts, neutral red, and crystal violet	Enteric gram-negative bacilli (see **colorplate 16**)
Eosin methylene blue agar (EMB)	Selective and differential	Lactose, eosin Y, and methylene blue	Enteric gram-negative bacilli
Hektoen enteric agar (HE)	Selective and differential	Lactose, sucrose, bile salts, ferric ammonium sulfate, sodium thiosulfate, bromthymol blue, acid fuchsin	*Salmonella* and *Shigella* species (enteric pathogens) (see **colorplate 17**)
Xylose-lysine-desoxycholate agar (XLD)	Selective and differential	Lactose, sucrose, xylose, phenol red	*Salmonella* and *Shigella*
Phenylethyl alcohol agar (PEA)	Selective	Phenylethyl alcohol (inhibits gram negatives)	Gram-positive bacteria
Colistin nalidixic acid agar (CNA)	Selective	Colistin and nalidixic acid (inhibit gram negatives)	Gram-positive bacteria
Modified Thayer-Martin agar (MTM)	Selective	Hemoglobin, growth factors, and antimicrobial agents	Pathogenic *Neisseria* species

Table 13.1 summarizes the most commonly used enriched, selective, and differential media, and identifies their purpose as primary media for the isolation of microorganisms. You should review the table before performing the exercise and as a guide to interpreting your results.

Purpose	To observe the response of a mixed bacterial flora in a clinical specimen to a battery of primary isolation media
Materials	Nutrient agar plates
	Blood agar plates
	MacConkey agar (MAC) plates
	Mannitol salt agar (MSA) plates
	Simulated fecal suspension, containing *Escherichia coli, Pseudomonas aeruginosa,* and *Staphylococcus epidermidis*
	Demonstration plates:
	Nutrient agar, blood agar, mannitol salt agar, and MacConkey agar plates, each divided into thirds, and one section on each streaked with a pure culture of *Staphylococcus aureus, Escherichia coli,* and *Pseudomonas aeruginosa*
	Wire inoculating loop
	Marking pen or pencil

Diagnostic Microbiology in Action

Procedures

1. Inoculate the simulated fecal specimen on nutrient agar, blood agar, MAC, and MSA plates. Streak each plate for isolation of colonies. Incubate at 35°C for 24 to 48 hours.
2. Make a Gram stain of the fecal suspension and examine it.
3. Examine the demonstration plates (do not open them) and record your observations.

Results

1. Demonstration plates: In the following chart, describe the appearance (e.g., shape, elevation [see fig. 2.1], pigment) of each organism on the culture media.

Medium	Staphylococcus aureus	Escherichia coli	Pseudomonas aeruginosa
Nutrient agar			
Blood agar			
Mannitol salt agar			
MacConkey agar			

2. Simulated fecal specimen cultures.

Medium	Gross Morphology of Each Colony Type	Gram-Stain Reaction and Microscopic Morphology of Each Colony Type	Presumptive Identification*

*Based on medium growth, colonial morphology, Gram-stain reaction, and microscopic morphology.

Questions

1. Define a *differential medium* and discuss its purpose.

2. Define a *selective medium* and describe its uses.

3. Why is MacConkey agar selective as well as differential?

4. Why is blood agar useful as a primary isolation medium?

5. How can one distinguish *E. coli* from *P. aeruginosa* on

   Nutrient agar? _____

   Blood agar? _____

   MAC agar? _____

6. What is the major difference between Modified Thayer-Martin (MTM) medium and chocolate agar? When would you use MTM medium rather than chocolate agar?

7. If you wanted to isolate *S. aureus* from a pus specimen containing a mixed flora, what medium would you choose to get results most rapidly? Why?

8. What is the value of making a Gram stain directly from a clinical specimen?

9. Why is aseptic technique important in the laboratory? In patient care?

10. Why is human blood no longer used in the preparation of culture media?

Some Metabolic Activities of Bacteria

Learning Objectives

After completing this exercise, students should be able to:

1. Discuss the basis of a simple carbohydrate fermentation reaction.
2. Describe how enzymes produced by various bacteria can be used for identification.
3. Specify the basis of the starch hydrolysis and indole reactions.
4. Explain the basis of the KIA test for identifying bacteria.
5. Name one indole-positive organism.

All forms of life must have a source of usable nutrients and energy in order to grow and carry on their vital processes. Energy is derived from the chemical breakdown of organic and inorganic compounds, except in organisms that rely on light for their energy source (e.g., by photosynthesis). The metabolic processes that microbes use to obtain energy for growth and reproduction are complex, but vital to their survival. In addition, tests for certain of these chemical reactions allow microbiologists to identify bacteria that are recovered in culture. Some of these activities and the enzymes that accomplish them are the subject of Exercise 15. In this exercise, you will learn how bacteria are identified by inoculating pure, isolated colonies into media that contain one or more specific biochemicals. The biochemical reactions that take place in the culture can then be determined by relatively simple indicator reagents, included in the medium or added to the culture later.

Some bacteria ferment simple carbohydrates, producing acids, alcohols, or gases as end products. Many different species are distinguished on the basis of the carbohydrates they do or do not attack, as well as by the nature of end products formed during fermentation. Still others break down more complex carbohydrates, such as in the hydrolysis of starch. The nature of products formed in amino acid metabolism also provides information as to the identification of bacterial species. The production of visible pigments distinguishes certain types of bacteria.

Working with pure cultures freshly isolated from clinical specimens, the microbiologist uses a carefully selected battery of special media to identify their characteristic biochemical properties. The microbes studied produce different reactions that aid in their identification because they produce different enzymes.

EXPERIMENT 14.1   Simple Carbohydrate Fermentations

Media for testing carbohydrate fermentation are often prepared as tubed broths, each tube containing a small inverted "fermentation" (or Durham) tube for trapping any gas formed when the broth is inoculated and incubated (see **colorplate 18**). Each broth contains essential nutrients, a specific carbohydrate, and may contain a color reagent to indicate a change in pH if acid is produced in the culture (the broth is adjusted to a neutral pH when prepared). Organisms that grow in the broth but do not ferment the carbohydrate produce no change in the color of the medium, and no gas is formed. Some organisms may produce acid products in fermenting the sugar, but no gas, whereas others may form both acid and gas. In some cases, organisms that do not ferment the carbohydrate use the protein nutrients in the broth, thereby producing alkaline end products, a result that is also evidenced by a change in indicator color.

Purpose	To distinguish bacterial species on the basis of simple carbohydrate fermentation
Materials	Tubed phenol red glucose broth
	Tubed phenol red lactose broth } with Durham tubes
	Tubed phenol red sucrose broth
	Slant cultures of *Escherichia coli, Serratia marcescens, Pseudomonas aeruginosa,* and *Proteus vulgaris*
	Wire inoculating loop
	Marking pen or pencil
	Test tube rack

Procedures

1. Each student or team of students should have four tubes of each carbohydrate broth (glucose, lactose, and sucrose), for a total of 12 tubes. Be certain the tubes are labeled with the name of the carbohydrate they contain and then label one of each set with the name of the four organisms to be tested.
2. Inoculate growth from each of the four cultures into the appropriately labeled tubes. Be certain no bubbles are inside the Durham tubes before inoculation.
3. Incubate at 35°C for 24 hours.

Results

Record your results in the following table. Use these symbols to indicate specific changes observed in the broths:

A = acid production
K = alkaline color change
N = neutral (no change in color)
G = gas formation

Name of Organism	Glucose	Lactose	Sucrose
Escherichia coli			
Serratia marcescens			
Pseudomonas aeruginosa			
Proteus vulgaris			

EXPERIMENT **14.2**   **Starch Hydrolysis**

Some microorganisms possess certain proteins known as enzymes, which split apart (hydrolyze) large organic molecules and then use the component parts in other metabolic activities. Starch is a complex polysaccharide that is hydrolyzed by some bacteria that produce the enzyme amylase. When iodine is added to the *intact* starch molecule, a blue-colored complex forms. If starch is hydrolyzed by amylase, however, it is broken down to simple sugars (glucose and maltose) that do not complex with iodine, and no color reaction is seen.

Figure 14.1 *Bacillus subtilis* colony on a culture medium containing starch. In this case, a large inoculum of the organism was placed in the center of the plate to form a "giant" colony after 24 hours of incubation. The culture plate has been flooded with a weak iodine solution, which reveals a zone of clearing around the colony (arrow). This zone represents the area where the starch has been hydrolyzed so that it is no longer available to react with the iodine solution. ©*Josephine A. Morello*

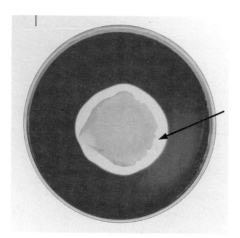

The medium for this test is a nutrient agar containing starch, prepared in a Petri plate. The organism to be tested is inoculated on the plate. When the culture has grown, the plate is flooded with Gram's iodine solution. The medium turns blue in all areas where the starch remains intact. The areas of medium surrounding organisms that have hydrolyzed the starch remain clear and colorless (see fig. 14.1).

Purpose	To distinguish bacterial species on the basis of starch hydrolysis
Materials	Starch agar plates
	Slant cultures of *Escherichia coli, Pseudomonas aeruginosa,* and *Bacillus subtilis*
	Gram's iodine solution
	Wire inoculating loop
	Marking pen or pencil

Procedures

1. Take one starch plate, invert it, and with your marking pen or pencil mark three triangular compartments on the back of the dish. Label one of the three compartments *E. coli;* a second, *P. aeruginosa;* and the third, *B. subtilis.*
2. Pick up a loopful of the *E. coli* culture and inoculate the agar heavily by rubbing the inoculum in a circle about the size of a quarter in the middle of the appropriate triangular area. Figure 14.1 shows how the growth will appear as a "giant" colony after incubation.
3. Repeat the procedure with the *P. aeruginosa* and *B. subtilis* cultures.
4. Incubate 24 to 48 hours at 35°C.
5. When the cultures have grown, drop Gram's iodine solution onto the plate until the entire surface is lightly covered.

Some Metabolic Activities of Bacteria

Results

Read and record your results in the table.

Name of Organism	Color Around Colony	Starch Hydrolysis (+ or −)
Escherichia coli		
Pseudomonas aeruginosa		
Bacillus subtilis		

EXPERIMENT **14.3**   **The Indole Test**

Some bacteria produce the enzyme tryptophanase, which degrades the amino acid tryptophan to produce pyruvic acid, ammonia, and indole. When bacteria are grown in a medium containing tryptophan, the presence of indole can be detected by adding a chemical solution, Kovac's reagent, to the culture. If indole is present, it immediately combines with the Kovac's reagent to produce a brilliant red color. If indole is not present, there is no color change and the reagent maintains its original color. As you will see in Exercises 23 and 24, the spot indole test, which is one method for performing this test, is used in a battery of different identification schemes to identify enteric bacteria, particularly *Escherichia coli*, a gram-negative bacillus commonly encountered in the clinical microbiology laboratory.

Purpose	To distinguish bacterial species by their production of indole
Materials	Tryptic soy agar plates with 5% sheep blood
	Slant cultures of *Escherichia coli*, *Proteus vulgaris*, and *Klebsiella pneumoniae*
	Kovac's reagent
	Sterile cotton-tipped swabs
	Wire or disposable inoculating loop
	Marking pen or pencil
	Test tube rack

Procedures

1. Invert a blood agar plate and mark three triangular compartments with your marking pen or pencil on the bottom of the plate. Label one compartment *E. coli*; a second, *P. vulgaris*; and the third, *K. pneumoniae*.
2. With your sterilized wire or disposable inoculating loop, pick up a loopful of the *E. coli* culture from the agar slant. Inoculate the blood agar plate by streaking the inoculum onto the appropriate triangular area of the blood agar plate.
3. Repeat this procedure with the *P. vulgaris* and *K. pneumoniae* slant cultures.
4. Incubate the plates for 24 to 48 hours at 35°C.
5. When the cultures are grown, moisten the tip of a sterile swab with the Kovac's reagent and touch it to a colony of *E. coli*.
6. Observe the swab for an immediate color change.
7. Discard the swab into the disinfectant container and record your results in the table under Results.
8. Using separate swabs, repeat this procedure with the *P. vulgaris* and *K. pneumoniae* cultures.

Results

Record your results in the table.

Organism	Indole Reaction (+ or −)
Escherichia coli	
Proteus vulgaris	
Klebsiella pneumoniae	

EXPERIMENT **14.4**   **Use of Kligler Iron Agar**

The use of Kligler iron agar (KIA) allows for detection of several biochemical reactions in a single tube of solid medium. KIA contains the carbohydrates glucose and lactose and a pH indicator that changes from red to yellow when acids are produced by fermentation of either of these simple sugars. In addition, KIA contains an iron salt (ferric ammonium citrate) and sodium thiosulfate. These two compounds permit detection of a bacterium's ability to break down sulfur-containing amino acids to produce hydrogen sulfide. The ferric ions combine with the released hydrogen sulfide to produce ferrous sulfide, resulting in the formation of a black precipitate in the medium.

In KIA medium, lactose is present in a concentration 10 times that of glucose. KIA is prepared as a slant, which sets up reaction chambers within the tube. The slant portion is the aerobic chamber exposed to atmospheric oxygen, and the butt, or deep portion, is the anaerobic chamber, because it is protected from air. The medium is inoculated with a well-isolated colony of a gram-negative organism growing on a culture plate. The colony is picked from the plate with an inoculating needle and the needle is "stabbed" into the KIA agar butt until the needle reaches the bottom of the tube. The needle is then withdrawn and used to inoculate the surface of the slant by moving the needle back and forth over the surface of the slant. The inoculated KIA slant is then incubated with the cap loose (this allows for air exchange) in a 35°C incubator for 24 hours, at which time the reactions are read.

Fermentation reactions are read on the slant and butt portions of the medium. Formation of an acid (yellow) butt and an alkaline (red) slant indicates that the organism ferments glucose but not lactose. Formation of an acid butt and an acid slant indicates that the organism ferments both glucose and lactose. If the butt and the slant remain red, the organism cannot ferment glucose or lactose, and it is referred to as a nonfermenter. Gas production is indicated by the presence of bubbles or a splitting or cracking of the agar medium. In addition, if the organism can produce hydrogen sulfide, blackening of the medium is observed. KIA slants are often used as preliminary screening tests to characterize certain groups of gram-negative bacteria, particularly those classified as enteric organisms. Another advantage of the KIA test is that it provides several biochemical test results in one medium, as is emphasized again in Exercises 23 and 24. **Colorplate 19** shows representative examples of KIA reactions for different groups of bacteria.

Purpose	To observe how a single medium can be used to determine multiple biochemical reactions of bacteria
Materials	Slants of KIA medium
	Slant cultures of *Escherichia coli*, *Proteus vulgaris*, and *Pseudomonas aeruginosa*
	Straight wire or disposable inoculating needle
	Marking pen or pencil
	Test tube rack

Procedures

1. Inoculate growth from each of the three slant cultures into separate labeled tubes of KIA medium. It is important to inoculate each KIA slant by stabbing the sterilized wire or disposable inoculating needle into the butt of the medium, then withdrawing the needle and inoculating the surface of the slant by moving the needle back and forth over the surface of the slant.
2. Incubate the KIA slants with their caps loose at 35°C for 18 to 24 hours.
3. Observe and record your results in the table.

Organism	Glucose Fermentation (+ or −)	Lactose Fermentation (+ or −)	Hydrogen Sulfide Production (+ or −)
Escherichia coli			
Proteus vulgaris			
Pseudomonas aeruginosa			

Questions

1. What is the color of phenol red at an acid pH?

2. What is the function of a Durham tube?

3. Why is iodine used to detect starch hydrolysis?

4. Name one indole-positive organism.

5. How is indole produced? How is it detected?

6. How is hydrogen sulfide demonstrated in KIA medium?

7. Name four biochemical reactions that can be observed in KIA medium.

8. Why is it necessary to incubate KIA slants with their caps loosened?

9. Why is it essential to have pure cultures for biochemical tests?

10. Could a pH-sensitive color indicator be used to reveal the presence of a contaminant in a fluid that should be sterile? Explain.

Activities of Bacterial Enzymes

Learning Objectives

After completing this exercise, students should be able to:
1. Discuss how enzymes function as catalysts.
2. Explain how enzymes work to break down compounds necessary for bacterial growth.
3. Describe the principles of the urease test, the catalase test, the oxidase test, the DNase test, and the deaminase test.
4. Discuss how biochemical tests can be used to identify bacteria.
5. Explain how color changes occur in each of the biochemical test media used in this exercise.

Enzymes are the most important chemical mediators of every living cell's activities. These enzymes are proteins that catalyze, or promote, the uptake and use of raw materials needed for synthesis of cellular components or for energy. Enzymes are also involved in the breakdown of unneeded substances or of metabolic side products that must be eliminated from the cell and returned to the environment.

As catalysts, enzymes promote changes only in very specific substances called *substrates*. Thus, in the previous exercise, the changes produced in simple carbohydrates and in starch substrates were brought about by different, specific enzymes. We have seen the activity of an enzyme with a different kind of outcome, the breakdown of an antimicrobial agent (see Experiment 12.3), but the principle is exactly the same. In the latter instance, the beta-lactamase enzyme penicillinase brought about a change in the substrate penicillin.

Because the activity of an enzyme is limited to a particular substrate, it follows that each bacterial cell must possess a large battery of different enzymes, each mediating a different metabolic process. They are identified in terms of the type of change produced in the substrate. In naming them, the suffix *-ase* is usually added to the name of the substrate affected. Thus, *urease* is an enzyme that degrades urea, *DNase* breaks down DNA, *penicillinase* inactivates penicillin, and so on.

In this exercise, we shall see how many bacterial enzymes are detected by performing laboratory tests and how their recognition in bacterial cultures leads to the identification of species.

EXPERIMENT 15.1   The Activity of Urease

Some bacteria split the urea molecule in two, releasing carbon dioxide and ammonia. This reaction, mediated by the enzyme urease, can be seen in a culture medium in which urea has been added as the substrate. Phenol red is also added as a pH indicator. When bacterial cells that produce urease are grown in this medium, urea is degraded, ammonia is released, and the pH becomes alkaline. This pH shift is detected by a change in the indicator color from orange-pink to dark pink (see **colorplate 20**).

Rapid urease production is characteristic of *Proteus* species, a few other enteric bacteria that at one time were classified in the *Proteus* genus, and a more recently discovered bacterium, *Helicobacter pylori*. *H. pylori* organisms are responsible for producing gastric ulcers and, in some cases, stomach cancer. By splitting urea with the release of alkaline ammonia, *H. pylori* can partially neutralize the strong acid in its gastric environment. Thus, while ammonia helps permit the organism's survival, its action, along with other *H. pylori* enzymes, is toxic for the cells lining the stomach. The result is damage to stomach tissue, which leads to ulcer disease. Whereas microbial enzymes are responsible for many activities that benefit microorganisms, humans, and the environment, as seen with *H. pylori*, they often play a role in the *pathogenesis* of infection. The urease test is useful in distinguishing urease-producing organisms from other bacteria that resemble them.

| | | Purpose | To observe the activity of urease and to distinguish bacteria that produce it from those that do not |

Purpose	To observe the activity of urease and to distinguish bacteria that produce it from those that do not
Materials	Tubes of urea broth or agar Slant cultures of *Escherichia coli* and *Proteus vulgaris* Wire inoculating loop Marking pen or pencil Test tube rack

Procedures

1. Inoculate a labeled tube of urea broth or agar with *E. coli,* and another with the *Proteus* culture.
2. Incubate the tubes at 35°C for 24 hours.

Results

Record your observations.

			Urease	
Name of Organism	Color of Urea Medium Before Culture	Color of Urea Medium After Culture	Positive	Negative
Escherichia coli				
Proteus vulgaris				

EXPERIMENT 15.2   The Activity of Catalase

Many bacteria produce the enzyme catalase, which breaks down hydrogen peroxide, releasing oxygen gas. The simple test for catalase can be very useful in distinguishing between organism groups. The hydrogen peroxide can be added directly to a slant culture or to bacteria smeared on a clean glass slide. The test should not be performed with organisms growing on a blood-containing medium because catalase is found in red blood cells, and thus, a falsely positive test result would be produced.

Purpose	To observe bacterial catalase activity
Materials	3% hydrogen peroxide Capillary pipettes Pipette bulb or other aspiration device Nutrient agar slant cultures of *Staphylococcus epidermidis* and *Enterococcus faecalis* Wire inoculating loop Clean glass slides Marking pen or pencil

Figure 15.1 Slide catalase test. *Staphylococcus epidermidis* on the left produces a strong positive catalase reaction. *Enterococcus faecalis* on the right (cloudy area in drop of hydrogen peroxide) is negative in the catalase test. ©*Josephine A. Morello*

Procedures

1. Divide a clean glass slide into two sections with your marking pen or pencil.
2. With a sterilized and cooled inoculating loop, pick up a small amount of the *Staphylococcus* culture from the nutrient agar slant. Smear the culture directly onto the left-hand side of the slide. The smear should be about the size of a pea.
3. Sterilize the loop again and smear a small amount of the *Enterococcus* culture on the right-hand side of the slide.
4. With the capillary pipette, place one drop of hydrogen peroxide over each smear. Be careful not to run the drops together. Observe the fluid over the smears for the appearance of gas bubbles (see fig. 15.1). Record the results in the chart. Discard the slide in a jar of disinfectant or a biohazard container.
5. Hold the slant culture of the *Staphylococcus* in an inclined position and pipette 5 to 10 drops of hydrogen peroxide onto the surface with the bacterial growth. Observe closely for the appearance of gas bubbles.
6. Repeat the procedure with the *Enterococcus* culture. Note whether oxygen is liberated and bubbling occurs.

Results

Record your observations and conclusions in this chart.

Organism	Slide Preparation		Tube Culture	
	Bubbling (+ or −)	Catalase (+ or −)	Bubbling (+ or −)	Catalase (+ or −)
Staphylococcus epidermidis				
Enterococcus faecalis				

EXPERIMENT 15.3   The Activity of Oxidase

The oxidase test detects bacteria that produce the enzyme cytochrome oxidase, which is involved in the transport of electrons in metabolic pathways of certain bacteria. The test is used commonly in clinical microbiology laboratories to identify certain genera of gram-negative bacteria. The oxidase test can be performed by using a filter paper method or a direct plate test. In the filter paper method, several drops of the oxidase reagent are first placed on a piece of filter paper. A colony of the organism to be tested is picked up with an applicator stick or a plastic disposable loop and rubbed onto the moistened area of the filter paper. A purple color will develop within 10 seconds in a positive reaction. A negative reaction produces no color change. Alternatively, the oxidase reagent may be dropped directly onto a colony growing on a culture plate. In a positive reaction, the colony first turns pink and then blue or purple (see **colorplate 13**). In a negative reaction, no color change occurs.

Purpose	To observe how the oxidase test is used for distinguishing between certain genera of gram-negative bacteria
Materials	Oxidase reagent
	Filter paper
	Cultures of *Escherichia coli, Klebsiella pneumoniae, Pseudomonas aeruginosa,* and a *Neisseria* sp. grown on blood or chocolate agar plates
	Applicator sticks, or wire or disposable inoculating loops

Procedures

A. Filter Paper Method

1. Moisten a piece of filter paper with several drops of oxidase reagent.
2. With an applicator stick or an inoculating loop, pick up several colonies of *E. coli* and rub this inoculum onto the moistened area of the filter paper. Observe the bacterial growth rubbed on the filter paper and note whether there is a color change to pink or purple within 10 seconds. Record your results in the table under Results.
3. Repeat Step 2 for the *K. pneumoniae, P. aeruginosa, and Neisseria* sp. cultures. Record all results in the table.

B. Direct Plate Method

1. Place several drops of the oxidase reagent directly onto a colony of *E. coli* that has been grown on a blood or chocolate agar plate. Observe for a change in color of the colony to pink or purple within 10 seconds. Record your results in the table.
2. Repeat step 1 with the *K. pneumoniae, P. aeruginosa, and Neisseria* sp. cultures and record all results in the table.

Results

	Filter Paper Method	Direct Plate Method
Organism	Oxidase (+ or −)	Oxidase (+ or −)
Escherichia coli		
Klebsiella pneumoniae		
Pseudomonas aeruginosa		
Neisseria sp.		

EXPERIMENT 15.4   The Activity of Deoxyribonuclease (DNase)

Some microorganisms secrete an enzyme that hydrolyzes the deoxyribonucleic acid (DNA) molecule. This can be demonstrated by inoculating a plated agar medium containing the substrate DNA with a culture of the organism that produces the enzyme. The uninoculated medium is opaque and remains so after the culture has grown. If the plate is then flooded with a solution of weak hydrochloric acid, a zone of clearing appears around colonies that have produced DNase. This clearing occurs because the large DNA molecule has been degraded by the enzyme, and the end products dissolve in the added acid. Intact DNA does not dissolve in weak acid but rather is precipitated by it; therefore, the medium in the rest of the plate, or around colonies that do *not* produce DNase, becomes more opaque. Another way to demonstrate breakdown of DNA by DNase is to flood the plate with 0.1% toluidine blue. Intact DNA will stain blue, and DNase-producing colonies will be surrounded by a pink zone (see **colorplate 21**).

Diagnostic Microbiology in Action

Purpose	To distinguish bacterial species that do and do not produce DNase
Materials	One DNA agar plate
	Dropping bottle containing 1 N HCl or 0.1% toluidine blue
	Slant cultures of *Escherichia coli* and *Serratia marcescens*
	Wire inoculating loop
	Marking pen or pencil

Procedures

1. Make a mark on the bottom of the DNA plate, dividing it in half.
2. Heavily inoculate one side of the plate with *Escherichia coli* by rubbing growth from the slant in a circular area about the size of a quarter (giant colony).
3. Inoculate the other side of the plate with *Serratia marcescens,* in the same manner.
4. Incubate the plate at 35°C for 24 hours.
5. Examine the plate for growth.
6. Drop 1 N HCl or 0.1% toluidine blue onto the agar surface until it is thinly covered with fluid.
7. Examine the areas around the growth on both sides of the plate for evidence of clearing or opacity, or a pink color if toluidine blue was used.

Results

Make a diagram of your observations.

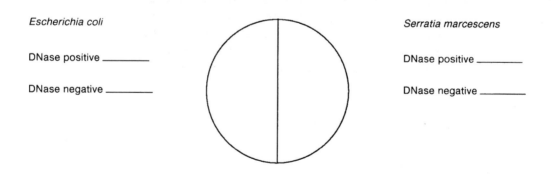

Escherichia coli

DNase positive _____

DNase negative _____

Serratia marcescens

DNase positive _____

DNase negative _____

EXPERIMENT **15.5**   **The Activity of a Deaminase**

Most bacteria possess a battery of enzymes that specifically break down individual amino acids. In the process, the amine group on the molecule is removed and the amino acid is degraded, the reaction being known as *deamination*. The deaminases that effect this type of change are named for the particular amino acid substrate for which they are specific. In this experiment, we will see the effects of a phenylalanine deaminase (PDase) produced by some bacteria.

When the amino acid phenylalanine is incorporated into a culture medium in which PDase-producing bacteria are growing, the substrate is degraded to phenylpyruvic acid. The reaction is made visible by adding ferric ions, which react with the newly produced acid to form a green compound. The appearance of a green color in a medium that was colorless when inoculated is evidence of the activity of the deaminase (see **colorplate 22**).

Purpose	To observe the activity of PDase and to distinguish bacteria that produce it from those that do not
Materials	Slants of phenylalanine agar
	Dropping bottle containing 10% ferric chloride
	Slant cultures of *Escherichia coli* and *Providencia stuartii*
	Wire inoculating loop
	Marking pen or pencil
	Test tube rack

Procedures

1. Inoculate each of the two cultures on a separate labeled slant of phenylalanine agar.
2. Incubate the new cultures at 35°C for 24 hours.
3. Examine the tubes for heavy growth. If it is adequate, run a few drops of 10% ferric chloride solution down the surface of each slant.
4. Observe the tubes for development of a green color.

Results

Organism	Color	PDase (+ or −)
Escherichia coli		
Providencia stuartii		

Questions

1. What is a catalyst?

2. Define an *enzyme* and a *substrate*. What is the value of enzyme tests in diagnostic microbiology?

3. What happens to urea in the presence of urease?

4. What is the substrate of the catalase reaction? Why are bubbles produced in a positive catalase test?

5. Why will a false-positive catalase test result if the organisms are tested on a medium containing blood?

6. What is the function of cytochrome oxidase?

7. Describe a positive DNase test.

8. What is a deaminase?

9. For each enzyme, indicate one bacterial species that produces it.

Urease _____

Catalase _____

Oxidase _____

Deoxyribonuclease _____

Phenylalanine deaminase _____

Antigen Immunoassays for Detection and Identification of Microorganisms

Learning Objectives

After completing this exercise, students should be able to:
1. Know which components of the microbial cell are antigenic.
2. Distinguish between a polyclonal and a monoclonal antibody.
3. List three types of antigen detection assays.
4. Describe the difference between direct and indirect enzyme immunoassays.

As you have seen in Exercises 13–15, the laboratory diagnosis of most infectious diseases involves the isolation in culture and subsequent identification of a microbial agent from a clinical specimen. Diagnoses can also be made by detecting antibodies in a patient's serum, as you will learn later in Exercise 33. Some of these traditional procedures are now being supplanted by new and more rapid methods that detect the presence of microorganisms or their products directly in patient specimens without the need for culture. In addition, these methods, which are often referred to as *nonculture methods,* can be used instead of biochemical tests to identify organisms that have already grown in culture. When used judiciously, these nonculture methods not only eliminate the need to perform culture, but also can be performed within minutes or hours. Thus, the time for reporting the result to the physician is shortened, and appropriate therapy can be administered to the patient sooner.

Clinical evaluations of nonculture technologies have shown that they are as reliable as, and in some cases, better than routine culture (i.e., more *sensitive* in detecting the microbe being sought). The result does not require growth of living, multiplying organisms but only detection of certain microbial cell structures or products. Another advantage is that these methods can detect infectious agents that, as yet, cannot be cultivated in the laboratory. An important example is the rotavirus, a common cause of infantile diarrhea that spreads rapidly in the hospital environment. Because this viral agent can now be detected directly in infant stool specimens by a rapid, nonculture method, its recognition helps prevent possible nursery-wide transmission.

Two types of nonculture methods are generally available. One type depends on detection of microbial antigens, a technology that has come into everyday use in clinical microbiology laboratories. Some examples are included in experiments you will perform in Part Three. The second type of nonculture method uses probes to detect microbial nucleic acids, sometimes in combination with techniques that greatly expand (amplify) small amounts of microbial DNA or RNA present in a patient specimen. These will be described in detail in the following exercise.

Because these nonculture methods detect the presence of specific molecules (i.e., microbial antigens or nucleic acids unique to a particular microorganism), they are collectively referred to as *molecular detection assays* or, more specifically, *antigen detection* or *nucleic acid assays,* respectively. The following discussion should aid your understanding of the principles of the antigen detection assays.

Antigen Detection Assays

All microorganisms contain a variety of different antigens whose composition is usually protein or carbohydrate in nature. Antigens may be components of the microbial cell wall, capsule, or intra- or extracellular enzymes. In humans, these antigens are recognized as foreign substances by the host immune system, which responds by producing specific protein molecules called *antibodies*. Antibodies bind specifically with the antigen that elicited their production. For example, the antigenic carbohydrate capsule of *Streptococcus pneumoniae* binds with its specific antibody in the quellung reaction (**colorplate 10**).

The use of an antibody to detect the presence of a specific microbial antigen is called an *immunoassay*. The sensitivity of immunoassays depends on the quality of the antibody preparation. In the early development of immunoassays, the antibody preparations, called *polyclonal antibodies,* were not pure enough to react only with a specific antigen (known as an antigenic determinant) on a specific microorganism. The result was often a *false-positive reaction*, in which a microorganism other than the one being tested for was detected because they shared a common antigenic determinant.

Through developments in immunology, a more specific type of antibody known as a *monoclonal* antibody can now be produced in large quantities. In contrast to the previously used polyclonal antibodies, monoclonal antibodies react with antigenic determinants that are unique to one microorganism and not shared by others. As a result, false-positive test reactions are greatly reduced and a wide variety of antigen-detection tests can now be performed in the clinical microbiology laboratory.

Many immunoassays are available, but three major types are in common use: *immunofluorescence, latex agglutination,* and *enzyme immunoassay* or *EIA*. For these immunoassays, monoclonal antibodies are labeled with (attached to) a "marker" molecule that provides a means of detecting whether an antigen-antibody reaction has taken place. Polyclonal antibody preparations are sometimes used for special purposes. The principles of these three assays are described briefly.

Immunofluorescence

In immunofluorescence assays, antibodies are labeled with a fluorescent dye called fluorescein. When the antibodies combine with their specific antigen in a preparation, the bright fluorescence can be visualized with a fluorescence microscope fitted with an ultraviolet illuminator and special filters. The test is performed by placing a smear of a clinical sample on a microscope slide and fixing it with a suitable reagent. In the simplest method, known as a *direct fluorescent antibody (DFA)* test, the fluorescein-labeled antibody preparation is applied directly to the specimen slide, which is then incubated, washed, and viewed under the fluorescence microscope. A positive test is indicated by the presence of brightly fluorescing organisms in the preparation (see **colorplates 23, 42, 43,** and **58**).

For the *indirect fluorescent antibody (IFA)* test, two antibody preparations are needed. The first, which is not labeled with the fluorescent dye, contains antibodies against the microbial agent we wish to detect. If the agent is present in the specimen smear, an antigen-antibody reaction occurs. To detect this combination, the second antibody, labeled with fluorescein, is applied to the preparation. This second antibody has been prepared to react with the first, unlabeled antibody. Again, a positive result is indicated by bright fluorescence under the microscope.

Figure 16.1 illustrates the principle of direct and indirect fluorescence assays. Fluorescent antibody tests are in widespread use to diagnose infections caused by a variety of microbial agents including bacteria, viruses, and protozoa.

Figure 16.1 (a) In the direct fluorescent antibody (DFA) test, an antibody specific for the microorganism sought is conjugated with the dye fluorescein. The antibody preparation is added to a specimen fixed to a slide. If the specific microorganism is present, the preparation will fluoresce when viewed under a fluorescence microscope. (b) In the indirect fluorescent antibody (IFA) test, the antibody specific for the microorganism is not conjugated with the dye, but will bind to the specific microorganism on the slide. A second antibody preparation, labeled with fluorescein, has been prepared to react with the first, unlabeled antibody and will fluoresce when viewed microscopically.

Source: Modified from J. M. Willey et al., *Prescott/Harley/Klein's Microbiology,* 7th ed., WCB/McGraw-Hill.

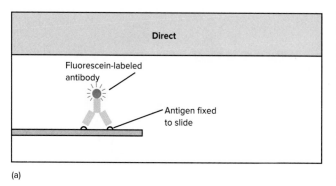

(a)

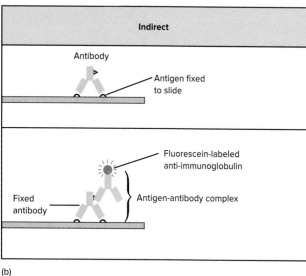

(b)

Latex Agglutination

In latex agglutination assays, antibodies are attached to latex (polystyrene) beads that serve as the marker for detecting the antigen-antibody interaction. Each latex particle is about 1 μm in diameter and can be charged with thousands of antibody molecules. Antibody-coated latex particles form a milky suspension, but when they are mixed with a preparation containing specific antigen, the resulting antigen-antibody complex results in visible clumping. Figure 16.2 illustrates the events associated with the latex particle agglutination reaction.

Latex agglutination tests are usually performed on a glass slide or a specially treated cardboard surface using small volumes of latex particles and liquid clinical sample. The reagent is mixed with the clinical sample by using a stirrer, and the slide is rocked by hand or rotated with a mechanical device for several minutes before being examined visually for clumping of the latex particles. **Colorplate 24** illustrates the appearance of positive and negative latex agglutination slide tests.

In clinical laboratories, latex agglutination tests are used to detect soluble microbial antigens directly in serum or cerebrospinal fluid specimens, or for identifying various types of bacteria recovered from culture plates. You will perform a latex agglutination test for *Staphylococcus aureus* in Experiment 19.1 and for *Streptococcus pyogenes* in Experiments 20.1 and 22.1.

Enzyme Immunoassay (EIA)

As in fluorescent antibody tests, the antibody in EIAs is conjugated with a marker that can be detected when an antigen-antibody reaction has taken place. In EIAs, the marker is an enzyme, typically alkaline phosphatase or horseradish peroxidase. These enzymes catalyze the breakdown of a colorless substrate to a colored end-product. To visualize the binding of antigen and the enzyme-linked antibody, the appropriate substrate for the enzyme must be added. A positive reaction results in the production of a colored end-product that can be detected visually or measured quantitatively by using a spectrophotometer.

Figure 16.2 Diagram of a latex reaction. (a) When latex particles that are coated with an antibody (e.g., group A *Streptococcus* antibody) react with a specific antigen (group A *Streptococcus* antigen), the particles join together to form clumps that agglutinate on the test slide (lower left corner of [a]). In the control test that is always run in parallel on the same slide, the same antigen is mixed with latex particles that are not coated with antibody; therefore, the particles remain in suspension and do not agglutinate. In diagram section (b), the antibody-coated particles do not react with the nonspecific antigen (e.g., group B *Streptococcus* antigen); therefore, no clumps are formed, and the test as well as the control suspension shows no agglutination (lower left corner of [b]).

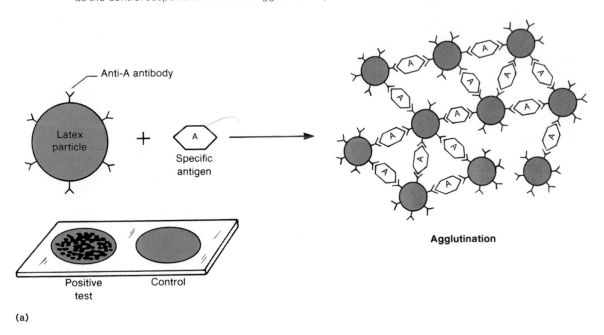

(a)

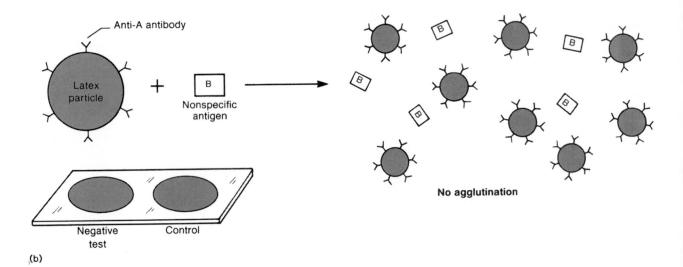

(b)

Diagnostic Microbiology in Action

Figure 16.4 An enzyme immunoassay kit for strep throat. The left side illustrates a negative test, with only the control (C) line positive. On the right, both the control (C) and patient specimen (T) are positive as illustrated by the appearance of two lines. *©Josephine A. Morello*

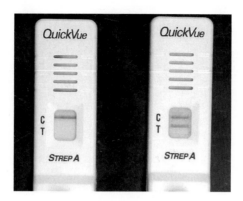

Questions

1. What are the two major types of nonculture technology used in many clinical microbiology laboratories?

2. List three advantages of nonculture methods over culture procedures for establishing a laboratory diagnosis of an infectious disease.

3. Name three types of immunoassays.

4. How does a direct immunoassay differ from an indirect assay?

5. What is a "marker" molecule?

Principles of Nucleic Acid Assays and Multiplex Syndrome Panel Testing for the Diagnosis of Infectious Disease

Learning Objectives

After completing this exercise, students should be able to:
1. List three types of nucleic acid detection assays.
2. Explain the difference between a probe assay and an amplification assay.
3. Distinguish between monoplex, biplex, and multiplex assays.
4. Name five infectious agents in each of the multiplex syndrome panels.
5. Explain the advantages of multiplex syndrome tests for diagnosing an infectious disease.

Nucleic Acid Detection Assays

Genes contain the genetic message (genotype) for all forms of life. The expression of the genotype results in the production of physical characteristics (phenotype) that make each life-form special and unique. All genes are made up of nucleic acids consisting of either DNA or RNA. Molecular biologists are now able to determine the order, or *sequence,* in which the nucleotides adenine, thymine, cytosine, and guanine (or uracil in RNA) occur in these nucleic acid molecules. Sequencing has revealed the entire set of genes (i.e., the *genome*) of many life-forms, including various microorganisms and even humans. By comparing the nucleotide sequence of genes among various life-forms, scientists can determine common regions of nucleic acids but, more importantly, they can determine the different nucleotide sequences that make a life-form special and unique.

With this knowledge, and the understanding that the nucleotide base adenine always bonds to thymine (in DNA) or uracil (in RNA), and guanine always bonds to cytosine, it is possible to synthesize in the laboratory a single-stranded sequence of nucleotides, known as a *probe* or *primer,* depending on the type of assay performed, which is complementary to a unique gene sequence in a specific life-form. When two single nucleic acid strands with complementary base sequences are placed together in solution, the nucleotide base pairs of each strand bond together to form a double-stranded molecule, called a duplex or hybrid. This *hybridization* reaction serves as the basis of two nucleic acid detection methods: *probe assays* and *amplification assays.*

Like the antigen detection assays described in Exercise 16, these nucleic acid detection assays have come into common use in many clinical microbiology laboratories, either to confirm the identity of a microorganism or to detect its presence directly in a clinical sample. The basic principles of probe and amplification assays are reviewed here.

Probe Assays

In a probe hybridization assay, one nucleic acid strand, known as the *probe,* will seek a complementary nucleic acid strand, the *target,* with which to combine. The probe is derived from a known microorganism or is manufactured synthetically, and the target is an unknown microorganism present in a clinical sample or isolated in culture. Like antigen detection assays, a *marker* or *reporter* molecule must be attached to the probe to determine whether the hybridization reaction has taken place (see fig. 17.1). Among the more popular reporter molecules are enzyme

Figure 17.1 Principles of nucleic acid hybridization. Identification of an unknown organism is established by positive hybridization (i.e., duplex formation) between a probe nucleic acid strand (from a known organism) and a target nucleic acid strand from the organism to be identified. Failure to hybridize indicates lack of homology between the probe and the target nucleic acid. Source: Reprinted from *Bailey & Scott's Diagnostic Microbiology,* 12th ed., by B.A. Forbes et al. Copyright © 2007 with permission from Elsevier.

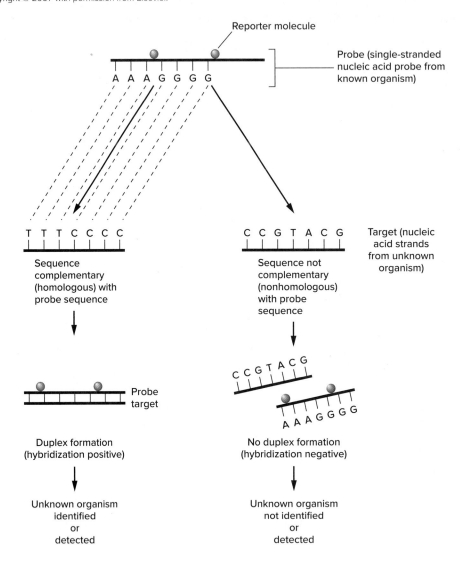

conjugates of alkaline phosphatase or horseradish peroxidase and chemiluminescent molecules such as acridinium. Chemiluminescent reporter molecules emit light that can be measured by using a special instrument called a luminometer. The amount of reporter molecule detected is directly proportional to the amount of hybridization that has occurred.

A positive hybridization reaction indicates that the unknown target is the same as the organism that served as the probe source. If no hybridization is detected, the target organism is not present in the sample (see fig. 17.1).

Single-stranded probe molecules can be composed of either DNA or RNA. Thus, double-stranded hybrids may be DNA-DNA, RNA-RNA, or DNA-RNA. Commercially available kits in which the probe molecule is DNA and the target molecule is a complementary strand of ribosomal RNA are popular because a microbial cell has several thousand copies of ribosomal RNA sequences but only one or two copies of DNA targets. As a result, these RNA-directed

Diagnostic Microbiology in Action

probes are far more sensitive for detecting low numbers of a microorganism in a sample than are DNA-directed probes.

Probe hybridization assays are used to confirm the identity of a wide variety of bacteria, fungi, viruses, and protozoa. Although they were once used commonly to detect the sexually transmitted bacterial pathogens *Chlamydia trachomatis* and *Neisseria gonorrhoeae* directly in clinical specimens, the more sensitive amplification assays have now gained popularity.

Amplification Assays

Often, too few bacteria may be present in a clinical specimen to be detected by a probe assay. To overcome this problem, a variety of methods referred to as *amplification* techniques have been devised. The most popular of these (for which Dr. Kary Mullis won the Nobel prize) is called the *polymerase chain reaction* or *PCR*. In this method, the specimen is heated to separate bacterial DNA strands, known bacterial primers are added to the mixture, and if they match the unknown single-stranded DNA, they combine (anneal) with it. Because the primers are shorter sequences than the original DNA strands, nucleotides and a heat-resistant Taq polymerase enzyme (originally isolated from a bacterium living in a hot spring in Wyoming's Yellowstone National Park) are added to the mixture to complete the formation of the double-stranded DNA. The new double-stranded DNA, referred to as an *amplicon,* is again separated, annealed with new primers, and extended with nucleotides in the presence of the polymerase enzyme. Each PCR step is carried out at a different temperature, which is automatically controlled by an instrument known as a *thermocycler.* The cycling is continued for up to 40 cycles, during which the original DNA sequences are increased or amplified a billionfold and, thus, many copies are available for detection. Probes labeled with reporter molecules that provide a chemiluminescent, fluorescent, or EIA-type colored signal are popular methods for amplicon detection. Figure 17.2 illustrates the steps in the PCR reaction.

The first PCR assays developed were essentially research methods. They involved the use of multiple manual steps; were tedious, time-consuming procedures that often required several days to complete; and had limited application for the clinical microbiology laboratory. Within the last few years, PCR technology has improved dramatically and become more user friendly, making it more suitable for diagnostic testing in the clinical laboratory. Automated thermocyclers that perform rapid thermocycling or "rapid cycling" combined with the continuous monitoring of the amplified product (amplicon) are now commercially available from several manufacturers. Because these rapid thermocycling instruments can complete the PCR assay in several hours, the test is commonly referred to as "rapid-cycle, real-time PCR," or simply "real-time PCR." Another type of PCR technology has now been developed, which performs the gene amplification reaction at a single temperature and thus eliminates the need for expensive equipment, such as a thermocycler. This new test is called "isothermal PCR."

Real-time PCR assays are available for the detection of a variety of infectious agents, and this technology is now widely accepted and recognized as the new "gold standard" (replacing culture) for the diagnosis of many infectious diseases. In addition to many conventional groups of bacteria, viruses, fungi, and protozoa that can be detected by this test, real-time PCR assays are particularly useful for the detection of infectious agents that cannot be cultured (e.g., hepatitis C virus), that are difficult to grow (e.g., *Mycobacterium leprae* and human metapneumovirus), or that, for safety reasons, the laboratory would not want to culture (e.g., variola virus or the severe acute respiratory syndrome (SARS)-related coronavirus). The availability of real-time PCR assays for rapid diagnosis can greatly impact patient care, direct the use of appropriate antimicrobial therapies, and reduce hospital stays and overall medical costs.

Other nucleic acid amplification assays have been developed for the rapid detection of microorganisms in clinical specimens, but their principle is similar to that of PCR. An amplification assay for detecting any infectious agent can be developed as long as the appropriate primer

Figure 17.2 Six cycles of a PCR reaction: after 40 cycles, the DNA sequences are amplified a billionfold. Source: Modified from K. P. Talaro and B. Chess, *Foundations in Microbiology,* 8th ed., McGraw-Hill.

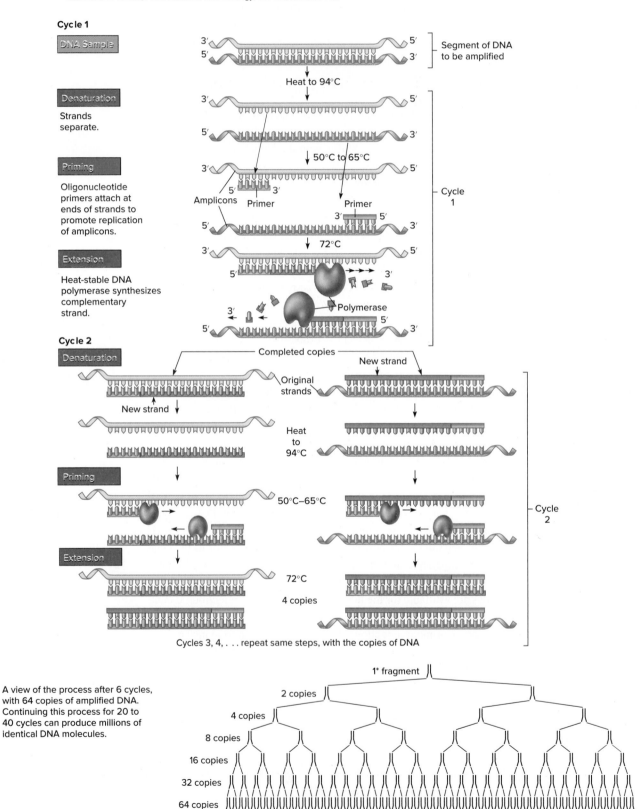

Cycle 1

DNA Sample — Segment of DNA to be amplified

Heat to 94°C

Denaturation
Strands separate.

50°C to 65°C

Priming
Oligonucleotide primers attach at ends of strands to promote replication of amplicons.

Amplicons Primer   Primer

72°C

Extension
Heat-stable DNA polymerase synthesizes complementary strand.

Polymerase

Cycle 1

Cycle 2

Denaturation
Completed copies

New strand

Original strands

New strand

Heat to 94°C

Priming
50°C–65°C

Extension
72°C
4 copies

Cycle 2

Cycles 3, 4, . . . repeat same steps, with the copies of DNA

A view of the process after 6 cycles, with **64 copies** of amplified DNA. Continuing this process for 20 to 40 cycles can produce millions of identical DNA molecules.

1ˢᵗ fragment

2 copies

4 copies

8 copies

16 copies

32 copies

64 copies

Diagnostic Microbiology in Action

sequence is available to begin the amplification reaction. Because these assays are so sensitive, laboratory personnel must take great care to avoid contaminating clinical samples with extraneous DNA that could be amplified and result in false-positive reactions and erroneous diagnoses.

The use of probe and amplification assays in the clinical microbiology laboratory are discussed further in other exercises of Part Three.

Multiplex Syndrome Panel Testing Using PCR Assays for the Diagnosis of Infectious Disease

Background

Over the last decade, the use of gene amplification assays, such as PCR, for the detection of various infectious agents directly in clinical specimens has become commonplace in clinical microbiology laboratories throughout the world. These molecular assays are considerably more reliable than conventional methods, and they provide final test results, usually within 1 to 2 hours of specimen receipt compared to days and sometimes weeks when cultural methods are used. Gene amplification assays have had their greatest application in the rapid diagnosis of certain respiratory viral infections, such as influenza A, influenza B, and respiratory syncytial virus infections, and some bacterial infections, such as whooping cough, caused by *Bordetella pertussis*. A major limitation of the use of these PCR assays is that they screen only for the presence of one or two microbes (called a monoplex or biplex PCR assay, respectively) in a single test, when, in fact, other microbial agents may be responsible for the infectious disease process.

To overcome this major limitation, multiplex PCR assays, called syndrome panels, which detect a large number of potential pathogens directly in a single specimen, have been developed. Currently, such panels are commercially available for the diagnosis of respiratory infections, gastrointestinal infections, central nervous system infections (meningitis and meningoencephalitis), and the common causes of septicemia (bloodstream infections). In addition, some of these assays can also detect the presence of antibiotic resistance genes that may be present in some bacteria. Like the monoplex and biplex PCR assays, with the proper instrumentation, the multiplex assays are relatively easy to perform, and the final results are available within 1 to 2 hours. The major advantage of the multiplex assays is that a single specimen type can be reliably tested to determine which of a number of microbial agents could be causing the patient's infectious disease. In addition, if the infection is caused by a bacterium and an antibiotic-resistance gene is detected, appropriate antibiotic therapy can be administered promptly.

It is not within the scope of this introductory laboratory manual to review in detail the technology for each of the multiplex assays currently in use. Instead, interested students are referred to the manufacturers' websites to learn more about these assays: BD Max™ (Becton-Dickinson Microbiology Systems, Cockeysville, MD); Verigene® (Nanosphere, North Chicago, IL); Luminex® (Luminex, Austin, TX); and BioFire FilmArray® (bioMérieux, Durham, NC).

As an example, the BioFire FilmArray® assay is used to illustrate the applications of the multiplex syndrome panel testing (fig. 17.3). The final results of each of the BioFire assays are available within 1 hour.

Respiratory Syndrome Panel

The Respiratory Syndrome Panel screens for 20 of the most common respiratory viral and bacterial pathogens that may present with nearly indistinguishable symptoms. The test is performed on a nasopharyngeal specimen. The rapid and reliable diagnosis of the patient's respiratory infection

Figure 17.3 The bioMerieux FilmArray system. The BioFire instrument shown in (a) interacts with the reagent pouch (b) to purify nucleic acids and amplify targeted nucleic acid sequences. In this way a large number of potential pathogens can be detected directly in a single specimen. ©*Verna Morton*

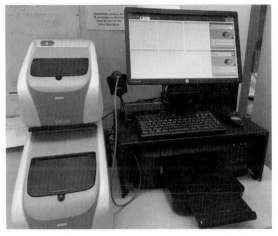

(a)

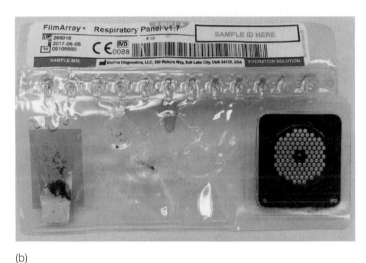

(b)

guides the health care provider to the choice of appropriate therapy, if indicated. The list of viral and respiratory pathogens detected in the multiplex panel follows:

VIRUSES

- Adenovirus
- Coronavirus HKU1
- Coronavirus NL63
- Coronavirus 229E
- Coronavirus OC43
- Human metapneumovirus
- Human rhinovirus/enterovirus
- Influenza A
- Influenza A/H1

- Influenza A/H3
- Influenza A/H1-2009
- Influenza B
- Parainfluenza virus 1
- Parainfluenza virus 2
- Parainfluenza virus 3
- Parainfluenza virus 4
- Respiratory syncytial virus

BACTERIA

- *Bordetella pertussis*
- *Chlamydophila pneumoniae*
- *Mycoplasma pneumoniae*

Gastrointestinal Syndrome Panel

The gastrointestinal panel screens for 22 bacterial, protozoan, and viral pathogens and is performed directly on a stool specimen. The prompt and accurate diagnosis of this infection can result in appropriate patient management and antimicrobial treatment, when indicated. The microbial agents detected in this assay include the following:

BACTERIA

- *Campylobacter* species (*jejuni, coli,* and *upsaliensis*)
- *Clostridium difficile* (toxin A/B)
- *Plesiomonas shigelloides*
- *Salmonella*
- *Yersinia enterocolitica*
- *Vibrio* species (*parahaemolyticus, vulnificus,* and *cholerae*)

DIARRHEAGENIC *E. COLI/SHIGELLA*
- Enteroaggregative *E. coli* (EAEC)
- Enteropathogenic *E. coli* (EPEC)
- Enterotoxigenic *E. coli* (ETEC) it/st
- Shiga-like toxin-producing *E. coli* (STEC) stx1/stx2
- *E. coli* O157
- *Shigella*/Enteroinvasive *E. coli* (EIEC)

PARASITES
- *Cryptosporidium*
- *Cyclospora cayetanensis*
- *Entamoeba histolytica*
- *Giardia lamblia*

VIRUSES
- Adenovirus F40/41
- Astrovirus
- Norovirus GI/GII
- Rotavirus A
- Sapovirus (I, II, IV, and V)

Blood Culture Identification Panel

The Blood Culture Identification Panel assay is performed directly on positive blood cultures and detects 24 of the most common gram-positive and gram-negative bacteria and yeast (*Candida* spp.) that cause septicemia. In addition, the assay detects the presence of three antibiotic resistance genes that may be harbored by some bacteria. Rapidly identifying the cause of septicemia and determining the presence or absence of bacterial resistance genes guide the physician in the choice of appropriate antibiotic therapy, which could impact patient outcome and survival. The various microbial agents that are detected by the Blood Culture Identification Panel are as follows:

GRAM-POSITIVE BACTERIA
- *Enterococcus*
- *Listeria monocytogenes*
- *Staphylococcus* species not *aureus*
- *Staphylococcus aureus*
- *Streptococcus* species
- *Streptococcus agalactiae*
- *Streptococcus pneumoniae*
- *Streptococcus pyogenes*

GRAM-NEGATIVE BACTERIA
- *Acinetobacter baumannii*
- *Haemophilus influenzae*
- *Neisseria meningitidis*
- *Pseudomonas aeruginosa*
- *Enterobacteriaceae*
- *Enterobacter cloacae* complex
- *Escherichia coli*
- *Klebsiella oxytoca*
- *Klebsiella pneumoniae*

- *Proteus*
- *Serratia marcescens*

YEAST
- *Candida albicans*
- *Candida glabrata*
- *Candida krusei*
- *Candida parapsilosis*
- *Candida tropicalis*

ANTIMICROBIAL RESISTANCE GENES
- mecA – methicillin resistance
- vanA/B – vancomycin resistance
- KPC – carbapenem resistance

Meningitis/Encephalitis Syndrome Panel

The Meningitis/Encephalitis Syndrome Panel is performed directly on spinal fluid and detects the presence of 14 of the most common bacteria, viruses, and yeast that cause life-threatening central nervous system infections. Early diagnosis of this infection can result in the prompt administration of appropriate antimicrobial therapy, which can impact patient survival and minimize possible complications of infection. The microbial agents detected by this assay are:

VIRUSES
- Cytomegalovirus (CMV)
- Enterovirus
- Herpes simplex virus 1 (HSV-1)
- Herpes simplex virus 2 (HSV-2)
- Human herpesvirus 6 (HHV-6)
- Human parechovirus
- Varicella zoster virus (VZV)

BACTERIA
- *Escherichia coli K1*
- *Haemophilus influenzae*
- *Listeria monocytogenes*
- *Neisseria meningitidis*
- *Streptococcus agalactiae*
- *Streptococcus pneumoniae*

YEAST
- *Cryptococcus neoformans/gattii*

Summary

Multiplex syndrome assays are being used increasingly in clinical microbiology laboratories throughout the United States. These multiplex assays allow for the detection of a large number of microbial agents that might be responsible for an infectious disease process. In addition, these assays can also detect antibiotic resistance genes that might be present in certain bacteria. The rapid availability of final test results within 1 to 2 hours of specimen receipt in the clinical microbiology laboratory results in early diagnosis and improved patient care.

Questions

1. What is a "reporter" molecule?

2. Name two major types of nucleic acid detection methods.

3. Why are nucleic acid amplification assays more sensitive than nucleic acid probe assays?

4. What important property of Taq polymerase has allowed its use in the PCR reaction?

5. Briefly define primer, amplicon, and thermocycler.

6. List four major advantages of the use of real-time PCR.

7. Define a monoplex PCR assay, a duplex PCR assay, and a multiplex PCR assay.

8. What is a major disadvantage of monoplex and duplex PCR assays?

9. List three advantages of using a multiplex PCR assay for the diagnosis of an infectious disease.

Principles and Applications of MALDI-TOF Mass Spectrometry for the Rapid Identification of Bacteria and Fungi

Learning Objectives

After completing this exercise, students should be able to:
1. Know what the acronym MALDI-TOF stands for.
2. Explain the basic operation of MALDI-TOF technology.
3. Describe the workflow of MALDI-TOF within the clinical microbiology laboratory.
4. List the advantages of using the MALDI-TOF method compared with conventional methods.

Background

For more than a century, traditional methods for the identification of bacteria and fungi recovered from clinical specimens have been based on examination of colony morphology and performance of biochemical tube tests. These methods continue to be used in many clinical microbiology laboratories, but they are time-consuming and labor-intensive. Importantly, the biochemical tube tests depend on subjective interpretations, and final results may not be available for 24 to 48 hours. These limitations have been largely eliminated by the use of automated instrument platforms that not only provide rapid and reliable bacterial identifications within 4 to 6 hours but also provide antimicrobial susceptibility test results for each organism tested.

Gene amplification assays, such as PCR, which is described in Exercise 17, are also becoming more widely used for the detection of bacteria directly in clinical specimens. However, gene amplification assays are expensive and often require highly trained personnel to perform the tests. As a result, the use of gene amplification assays is often limited to larger hospital and reference laboratories.

Principles of MALDI-TOF MS

MALDI-TOF MS is an acronym for **M**atrix-**A**ssisted **L**aser **D**esorption **I**onization-**T**ime **O**f **F**light **M**ass **S**pectrometry. The technology has been incorporated into an automated instrument platform, which provides a rapid, reliable, and inexpensive alternative method for the identification of bacteria and fungi.

As shown in figure 18.1, the MALDI-TOF MS instrument is composed of three basic components: an ionization chamber, a mass analyzer, and an ion detector. An unknown bacterial or fungal isolate is spotted in a sample well on a target slide and overlaid with a chemical matrix. After drying, the slide is placed in the ionization chamber. Once the chamber door is closed, the spot on the slide is pulsed by a laser beam. The energy from the laser beam serves two functions: it ionizes the individual microbial protein molecules into positively charged ions and transforms the proteins from a solid phase into a gas, a process called desorption. The vaporized sample is then directed into and accelerated through the mass analyzer, which separates the positively

Figure 18.1 Instrument modules.

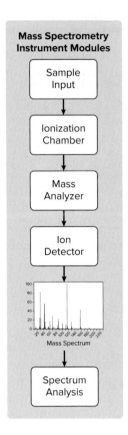

charged ions on the basis of their mass-to-charge ratio. Because the speed at which the ions travel depends on their mass-to-charge ratio, those with smaller ratios travel faster than those with larger ratios.

Upon emerging from the mass analyzer, the ionized particles are captured and their time-of-flight is measured by an ion detector. The mass spectrum generated is essentially a unique "protein" fingerprint of the unknown bacterium or fungus. Based on the protein spectral profile, the unknown microorganism is identified by computerized comparison of the unknown spectrum to a database of reference spectra from more than 22,000 previously well-characterized bacteria and fungi. Figure 18.2 illustrates the basic features of the MALDI-TOF instrument and technology.

Applications of MALDI-TOF MS

MALDI-TOF MS technology can be used to identify aerobic and anaerobic bacteria, mycobacteria, fungi, and yeast. Because of its ability to provide more rapid and reliable results, an increasing number of clinical microbiology laboratories are abandoning the use of conventional identification methods and adopting MALDI-TOF MS technology. In addition, the MALDI-TOF MS instrumentation is less expensive to use than the traditional biochemical tube and automated methods.

Figure 18.2 Basic features of the MALDI-TOF MS instrument and technology.

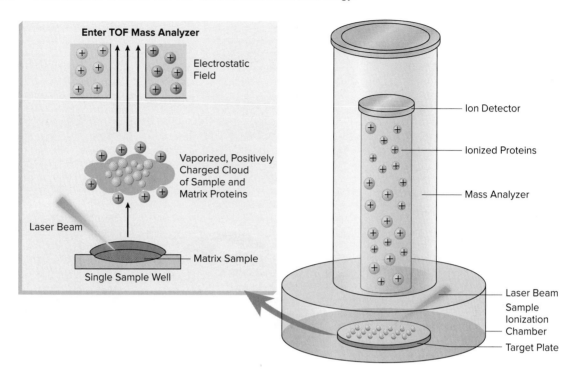

The standard workflow that many laboratories have adopted for the MALDI-TOF MS technology is shown in figure 18.3. Once a microorganism has grown on a culture plate, a small portion of a colony is transferred to a sample well on the target slide. The sample is then overlaid with a liquid matrix containing formic acid and allowed to air-dry. The target slide is placed in the sample ionization chamber. The chamber door is sealed and the chamber is evacuated. A laser beam is pulsed onto the slide and the MALDI-TOF MS process proceeds as described in the previous section. Once the procedure is completed, a final electronic and/or hard-copy report is generated, which can be transmitted or delivered to the hospital or doctor's office.

The MALDI-TOF MS target slide can accommodate up to 48 unknown isolates. Once the slide is placed in the instrument, identification takes approximately 15 to 20 seconds per microorganism. The labor and material costs for performing one MALDI-TOF MS identification are less than $0.50. In short, compared with other methods, the introduction of the MALDI-TOF MS technology into the clinical microbiology laboratory has resulted in more rapid and reliable identifications at considerably less expense.

Figure 18.3 MALDI-TOF MS laboratory workflow.

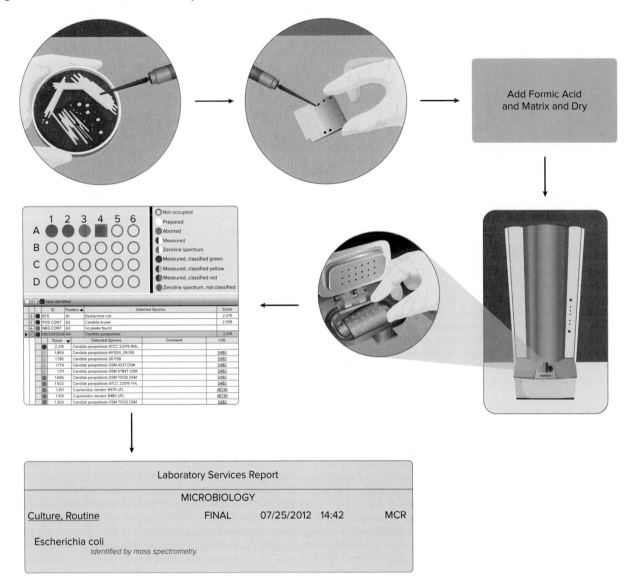

Questions

1. What does MALDI-TOF MS stand for?

2. Name the three basic components of a MALDI-TOF MS instrument.

3. Name four groups of microorganisms that can be identified by MALDI-TOF MS.

4. Describe the basic principle of operation for MALDI-TOF MS.

5. How long does it take for microbial identification by a conventional tube method compared with an automated instrument? Compared with MALDI-TOF MS instrumentation?

6. Identify five advantages of using the MALDI-TOF identification system versus conventional biochemical tube tests.

Staphylococci

Learning Objectives

After completing this exercise, students should be able to:
1. Describe a biochemical test to differentiate *Staphylococcus* from *Streptococcus* species.
2. Discuss selective media for isolating and differentiating *Staphylococcus* species.
3. List two diseases caused by *Staphylococcus* species.
4. Name the test that differentiates *Staphylococcus aureus* from most other *Staphylococcus* species.
5. Specify body sites where *Staphylococcus* species are found as part of the normal flora.

Staphylococci are ubiquitous in our environment and in the normal flora of our bodies. They are particularly numerous on skin and in the upper respiratory tract, including the anterior nares and pharyngeal surfaces. Some are also associated with human infectious diseases.

Staphylococci are gram-positive cocci, characteristically arranged in irregular clusters like grapes (see **colorplate 1**). They are hardy, facultatively anaerobic organisms that grow well on most nutrient media. There are three principal clinically important species: *Staphylococcus epidermidis, Staphylococcus saprophyticus,* and *Staphylococcus aureus. S. epidermidis,* as its name implies, is the most frequent inhabitant of human surfaces, including skin and mucous membranes. It is not usually pathogenic, but it may cause serious infections if it has an opportunity to enter past surface barriers, such as in cardiac surgery patients or those with indwelling intravenous catheters. *S. saprophyticus* has been implicated in acute urinary tract infections in young women between 16 to 25 years of age. It has not been found among the normal flora and is not yet confirmed to cause other types of infection. It is included in this exercise for completeness. Like *S. epidermidis, S. aureus* is often found among the normal flora of healthy persons, but in contrast, most staphylococcal disease is caused by strains of this species.

S. aureus strains produce a number of toxins and enzymes that can exert harmful effects on the cells of the infected host. Their *hemolysins* can destroy red blood cells. The enzyme *coagulase* coagulates plasma, but its exact role in staphylococcal infection is not yet known. *Leukocidin* is a staphylococcal toxin that destroys leukocytes. *Hyaluronidase* is an enzyme that acts on hyaluronic acid, a substrate that is a structural component of connective tissue. Its activity in a local area of infection breaks down the tissue and permits the staphylococci to penetrate more deeply; hence, it is called "spreading factor." (Some streptococci also produce hyaluronidase.) *Staphylokinase* can dissolve fibrin clots, thus enhancing the invasiveness of organisms that would otherwise be walled off by the body's fibrinous reactions. An *enterotoxin* is elaborated by some strains of *S. aureus*. If these are multiplying in contaminated food, the enterotoxin they produce can be responsible for severe gastroenteritis or staphylococcal food poisoning. Some strains produce toxic shock syndrome (TSS) by elaborating a toxin referred to as TSST-1. This disease is seen primarily in menstruating women who use highly absorbent tampons. *S. aureus* colonizing the vaginal tract multiplies there and releases TSST-1, causing a variety of symptoms including shock and a rash. TSS has also been documented in children, men, and nonmenstruating women who have a focus of infection at nongenital sites. Strains of *S. epidermidis* and *S. saprophyticus* do not produce these toxic substances.

Common skin infections caused by *S. aureus* include pimples, furuncles (boils), carbuncles, and impetigo. Serious systemic (deep tissue) infections that result from *S. aureus* invasion include pneumonia, pyelonephritis, osteomyelitis, meningitis, and endocarditis. In addition to pneumonia, *S. aureus* may also produce infections of the sinuses (sinusitis) and middle ear (otitis media).

EXPERIMENT 19.1   Isolation and Identification of Staphylococci

The laboratory diagnosis of staphylococcal disease is made by identifying the organism (usually *S. aureus*) in a clinical specimen representing the site of infection (pus from a skin lesion, sputum when pneumonia is suspected, urine, spinal fluid, or blood). It should be remembered that *S. aureus* and *S. epidermidis* may be harmless when present on superficial tissues. Special care must be taken not to contaminate the specimen with normal flora, and laboratory results must be interpreted in the light of the patient's clinical symptoms. The principal features by which staphylococci are recognized and distinguished from each other in the laboratory include colonial appearance on blood agar (especially hemolytic activity), coagulase activity (see **colorplate 26**), reaction to the carbohydrate mannitol, and susceptibility to the antimicrobial agent novobiocin (see **colorplate 27**). Also, a rapid latex agglutination test is available for identifying *S. aureus* from characteristic colonies growing on agar media. The antibody-coated latex beads react with two surface proteins typically found on *S. aureus* strains. One is a type of coagulase bound to the staphylococcal surface, and the other is a surface protein known as protein A (see fig. 19.1). The species are indistinguishable microscopically. On blood agar, *S. aureus* usually displays a light to golden yellow pigment (hence, its name), whereas *S. epidermidis* has a white pigment and *S. saprophyticus* either a bright yellow or white pigment. However, pigmentation is not always a reliable characteristic for distinguishing the species. On blood agar, *S. aureus* is usually, but not always, beta-hemolytic; *S. epidermidis and S. saprophyticus* are almost always nonhemolytic. *S. aureus* is, by definition, coagulase positive; *S. epidermidis* and *S. saprophyticus* are coagulase negative. Other *Staphylococcus* species that are found on skin but seldom cause disease are also coagulase negative. As a group, these species, along with *S. epidermidis* and *S. saprophyticus,* are referred to as *coagulase-negative staphylococci,* abbreviated CoNS. *S. aureus* is further distinguished by its ability to ferment mannitol, and *S. saprophyticus* by its resistance to low concentrations of novobiocin (see **colorplate 27**). Table 19.1 summarizes the major characteristics differentiating *Staphylococcus* species.

Because specimens from the mucous membranes or skin may contain a mixed normal flora as well as the pathogenic staphylococci being sought, the use of a selective, differential medium in the primary isolation battery can be very helpful (see table 13.1). Mannitol salt agar is such a medium. It contains a high concentration of salt that inhibits gram-positive cocci other than staphylococci and many other organisms as well. It also contains mannitol and an indicator to differentiate *S. aureus* strains from coagulase-negative staphylococci growing on it. A blood agar plate is also essential for demonstrating hemolytic organisms. Some streptococci, as well as many strains of *S. aureus,* are beta-hemolytic, so they can be distinguished promptly. Aside from microscopic morphology, the simplest, most rapid distinction can be made with the catalase test, for all streptococci are catalase negative, whereas all staphylococci are catalase positive.

S. aureus is carried by a large segment of the population as a member of the normal flora on the skin and mucous membranes, particularly the anterior nares. It causes disease primarily in individuals with lowered resistance, especially hospitalized patients. In the hospital, *S. aureus* is a major cause of nosocomial infections transmitted from hospital personnel or the

Figure 19.1   A rapid latex agglutination test for identifying *Staphylococcus aureus*. The top left and center wells are the positive and negative controls, respectively. The top right well is the positive reaction of the patient's isolate. *©Josephine A. Morello*

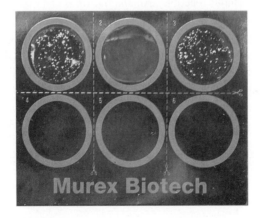

Diagnostic Microbiology in Action

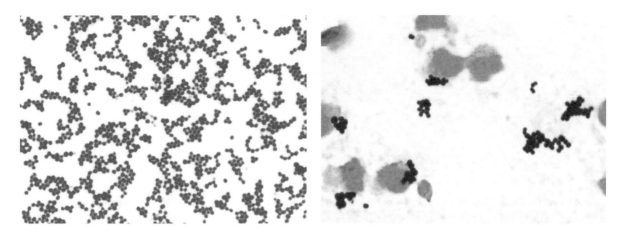

Plate 1 *Staphylococcus aureus* in a Gram-stained smear from a colony growing on agar medium (left) and from the sputum of a patient with staphylococcal pneumonia (right). The organisms are gram-positive spheres, primarily in grapelike clusters. The pink cells in the right-hand photo are neutrophils. ©Josephine A. Morello

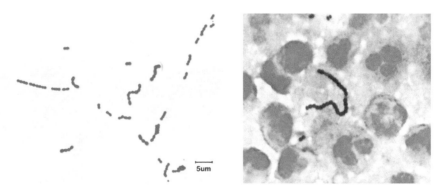

Plate 2 *Streptococcus pyogenes* in Gram-stained smears. The characteristic long chains of gram-positive cocci growing in broth are seen (left). Left: ©Gary E. Kaiser, Ph.D. In a smear from an abscess (right), the organism is seen primarily as long chains of gram-positive cocci, suggesting infection caused by a streptococcus. Right: ©Josephine A. Morello

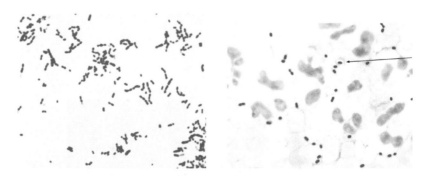

Plate 3 *Streptococcus pneumoniae* in Gram-stained smears. The organisms from a colony growing on agar medium (left) are gram positive and lancet shaped and appear in pairs and short chains. In a Gram-stained smear of cerebrospinal fluid from a patient with pneumococcal meningitis (right), the organisms are mostly diplococci. The capsule (arrow) can be seen around some bacteria, outlined by the pink proteinaceous material of the fluid. ©Josephine A. Morello

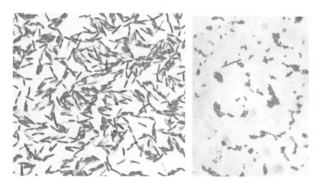

Plate 4 Gram-negative bacilli (*Klebsiella pneumoniae*) in a Gram-stained smear from an agar colony (left) and a patient's blood culture (right). In the blood specimen, the organisms are *pleomorphic,* varying in length from coccobacillary to filamentous.
Left: ©Josephine A. Morello; Right: ©Neal R. Chamberlain, Ph.D.

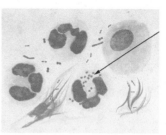

Plate 5 *Neisseria gonorrhoeae* in a Gram-stained smear from a female cervical exudate appear as gram-negative, bean-shaped diplococci, most of which are intracellular in a neutrophil (arrow).
Source: Dr. Norman Jacobs, CDC

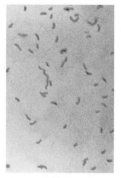

Plate 6 Curved, spiral, gram-negative bacilli (*Campylobacter jejuni*) in a Gram-stained smear from a culture. Some bacteria line up to form spirilla-like chains.
Source: CDC/Mediscan

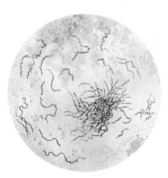

Plate 7 Spirochetes (*Treponema pallidum*) appear black in a skin preparation stained with a silver stain. Source: CDC

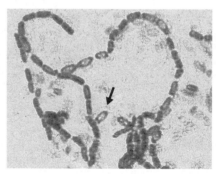
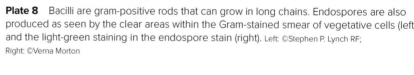

Plate 8 Bacilli are gram-positive rods that can grow in long chains. Endospores are also produced as seen by the clear areas within the Gram-stained smear of vegetative cells (left) and the light-green staining in the endospore stain (right). Left: ©Stephen P. Lynch RF; Right: ©Verna Morton

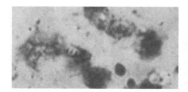

Plate 9 *Mycobacterium tuberculosis* in an acid-fast stain of sputum (left). The acid-fast bacilli appear as red, beaded rods against a blue background (×1,000). When stained with a rhodamine-auramine fluorescent dye, the acid-fast bacilli fluoresce brightly against a dark background (right, ×400). ©Josephine A. Morello

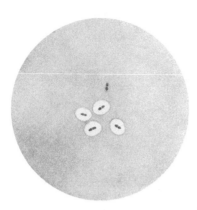

Plate 10 The quellung reaction. The halo around the cells is the pneumococcal capsule, which appears to swell when the cells are treated with pneumococcal antiserum. Source: CDC

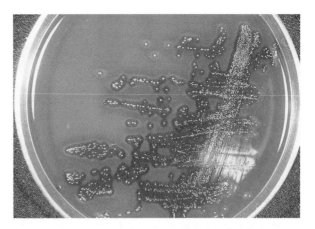

Plate 11 *Streptococcus pyogenes* growing on a blood agar plate. The clear areas of red blood cell lysis (beta hemolysis) surrounding the bacterial colonies are caused by the enzyme streptolysin, which is produced by the streptococci. ©Dr. E.J. Bottone

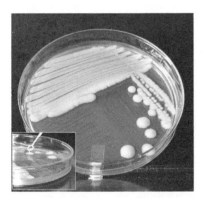

Plate 12 The large capsule of *Klebsiella pneumoniae* gives a mucoid appearance to colonies growing on agar plates. When growth is lifted with an inoculating loop, the capsule forms a sticky string (inset red arrow). ©Neal R. Chamberlain, Ph.D.

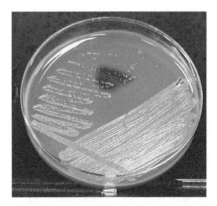

Plate 13 *Neisseria gonorrhoeae* colonies on chocolate agar. The oxidase-positive organisms become deep purple when a drop of oxidase reagent is added to an area of the plate. ©Neal R. Chamberlain, Ph.D.

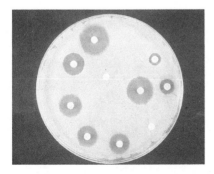

Plate 14 A disk agar diffusion antimicrobial susceptibility test. If the clear zones of growth inhibition around disks are of a certain diameter, the organism is susceptible to the antimicrobial agent in the disk. ©Josephine A. Morello

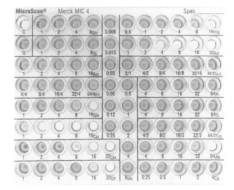

Plate 15 A microdilution susceptibility test of *Pseudomonas aeruginosa*. The green color signifies growth of the organism with production of its soluble green pigment. Growth occurs in the wells containing concentrations of antimicrobial agents to which the organism is resistant. ©Josephine A. Morello

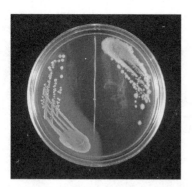

Plate 16 A MacConkey agar plate with *Escherichia coli* (pink, lactose-fermenting colonies) growing on the left-hand side and a *Salmonella* sp. (colorless, lactose nonfermenting colonies) on the right.
©Dr. E.J. Bottone

Plate 17 *Escherichia coli* (left) and *Salmonella* sp. (right) on a Hektoen enteric agar plate. The lactose-fermenting *E. coli* colonies appear yellow, whereas the *Salmonella* colonies appear black because of hydrogen sulfide production. Compare with reactions on colorplate 16 and note how selective and differential media display different organism characteristics. ©Josephine A. Morello

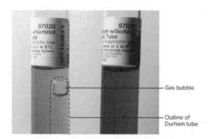

Gas bubble

Outline of
Durham tube

Plate 18 As seen on the left, when a Durham tube is placed inside a broth tube, gas produced by the fermentation of the carbohydrate in the medium can be seen as a bubble in the Durham tube. Acid production in the tube is detected by a lowering of the pH of the medium, resulting in a yellow color of the pH indicator. The tube on the right shows growth but no acid or gas production. ©Kenneth Van Horn

Plate 19 Kligler iron agar slants test for fermentation of glucose and lactose and the production of gas and hydrogen sulfide. The organism on the left ferments both glucose and lactose with gas production (bubbles in medium). The organism in the middle ferments glucose (yellow butt) but not lactose (pink slant). The organism on the right ferments glucose (with gas production) but not lactose, and blackens the agar as a result of hydrogen sulfide production. Reactions are similar on TSI slants. ©Josephine A. Morello

Plate 20 Urease test. The organism on the right produces the enzyme urease, which imparts the bright pink alkaline reaction to the urea agar slant.
©Verna Morton

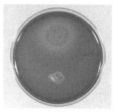

Plate 21 DNase test. When a plate containing DNA is flooded with toluidine blue, the colony of deoxyribonuclease-producing organisms (top) and the surrounding area of hydrolyzed DNA become pink. ©Josephine A. Morello

Plate 22 Phenylalanine deaminase (PDase) test. The PDase-producing *Providencia stuartii* (left) hydrolyzes phenylalanine in the culture medium. After ferric chloride is added to the slant, the green positive reaction appears. In the middle is the PDase-negative *Escherichia coli*; the tube on the right is uninoculated. ©Josephine A. Morello

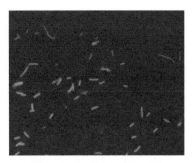

Plate 23 Fluorescent antibody preparation of *Legionella pneumophila* viewed microscopically with an ultraviolet light source. After the patient specimen is treated with an antibody conjugated with a fluorescent dye, the brightly fluorescing bacilli are easily visible against the dark background. ©Josephine A. Morello

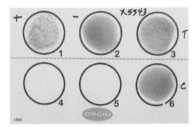

Plate 24 Latex agglutination reaction. Antibody-coated latex particles have been mixed with the positive (well 1) and negative (well 2) controls and the organism isolated from the patient (well 3). The fine granular clumps in the positive control and patient wells represent the positive reaction of agglutinated latex particles. Well 6 is an additional negative control. ©Cynthia Phillips; ©Scott Matushek

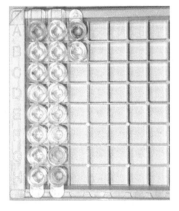

Plate 25 A direct enzyme immunoassay for *Clostridium difficile* toxin. The wells have been coated with antibody against the toxin and suspensions of patient fecal specimens added. The first well in row A and the second wells in rows G and H are positive, whereas the second well in row E shows a weakly positive reaction. In the third column, positive (row A) and negative (row B) controls are shown. Refer to figure 16.3 for details of the test. ©Josephine A. Morello

Plate 26 Coagulase test. The tube of plasma on the right was inoculated with *Staphylococcus aureus*. A solid clot has formed in this tube in comparison to the still liquid plasma in the uninoculated tube on the left. Both tubes are slanted to emphasize the differences. ©Dr. E.J. Bottone

Plate 27 Novobiocin disk test for differentiating two coagulase-negative species of staphylococci: *Staphylococcus saprophyticus* (left) and *Staphylococcus epidermidis* (right). The zone of inhibition around *S. saprophyticus* is less than 16 mm, which identifies this species by its resistance to the antibiotic. ©Josephine A. Morello

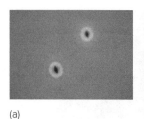  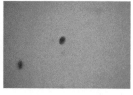

(a) (b) (c)

Plate 28 Subsurface colonies of alpha- (left), beta- (center), and nonhemolytic (right) streptococci. Note many intact red blood cells and a greenish color around the alpha-hemolytic colonies. Hemolysins produced by beta-hemolytic colonies have completely destroyed surrounding red blood cells. Nonhemolytic organisms produce no change in the red blood cells. ©Josephine A. Morello

Plate 29 Bacitracin test. The organism on the left is identified presumptively as *Streptococcus pyogenes* (group A), because it shows a zone of growth inhibition around the bacitracin disk. The bacitracin-resistant organism on the right is a beta-hemolytic streptococcus other than group A. ©Neal R. Chamberlain, Ph.D.

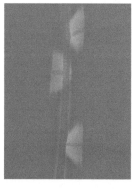

Plate 30 CAMP test. When a group B streptococcus (*Streptococcus agalactiae*) is streaked at right angles to a hemolytic *Staphylococcus aureus* (long straight streak down middle of plate), areas of synergistic hemolysis in the shape of a beta-hemolytic arrow are formed. ©Dr. E.J. Bottone

Plate 31 Optochin test. A zone of inhibition forming around a disk containing optochin (P disk) identifies this organism presumptively as *Streptococcus pneumoniae.* ©Josephine A. Morello

Plate 32 Esculin reaction. The *Enterococcus* sp. on the left has hydrolyzed esculin with resulting blackening of the medium. The *Streptococcus* sp. on the right does not hydrolyze esculin. ©Josephine A. Morello

Plate 33 PYR test. The appearance of a red color at the completion of the test indicates that the organism on the right is PYR positive. ©Josephine A. Morello

Plate 34 Satellite test. Colonies of *Haemophilus influenzae,* which requires both X and V factors, grow only around a *Staphylococcus aureus* streak on a blood agar plate. The blood provides the needed X factor (hemin) and the staphylococcus, V factor (a coenzyme, NAD). ©Verna Morton

Plate 35 Haemophilus ID Quad Plate inoculated with *Haemophilus influenzae.* The organism grows only on the top two quadrants, which contain media supplemented with X and V factors (left) and 5% blood and V factor (right).
©Josephine A. Morello

(a)

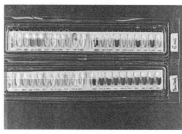
(b)

Plate 36 Rapid bacterial identification. In the Enterotube II (left) and API strips (right), many reactions are tested simultaneously allowing definitive organism identification within 24 hours. The top API strip shows typical positive reactions for an *E. coli* strain. The bottom (control) strip is rehydrated with saline only. (a) ©Catalin Rusnac/123RF RF; (b) ©Josephine A. Morello

Plate 37 Phenol red agar slants containing glucose, maltose, sucrose, and fructose inoculated with oxidase-positive, gram-negative diplococci. Only the first tube (glucose) shows a positive reaction, indicating the organism is *Neisseria gonorrhoeae.*
©Josephine A. Morello

Plate 38 The JEMBEC plate is used primarily when genital specimens for culture must be transported long distances to the microbiology laboratory. After the white CO_2-generating tablet (top right) is placed in the well in the rectangular culture plate, the plate is sealed in the plastic zip-lock bag. CO_2 accumulates, providing the appropriate atmosphere for growth of *Neisseria gonorrhoeae.* This plate is uninoculated. ©Josephine A. Morello

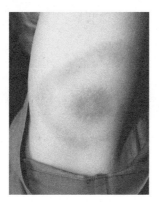

Plate 39 Clinical appearance of "bull's-eye" or erythema migrans skin lesion following a tick bite in a patient who developed Lyme disease.
Source: CDC Public Health Image Library

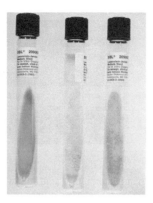

Plate 40 Growth of *Mycobacterium* spp. on Lowenstein-Jensen slants. The green tube on the left is uninoculated. The tube in the center has the characteristic dry, heaped, and rough growth of *M. tuberculosis*. The tube on the right shows the yellow pigmented growth of the photochromogen, *M. kansasii.* ©Josephine A. Morello

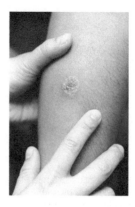

Plate 41 A positive PPD skin test, indicating the person has been infected with *Mycobacterium tuberculosis*. ©Mark Thomas/Science Source

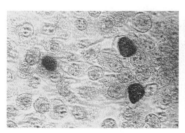
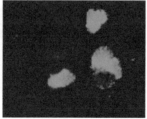

Plate 42 Inclusions of *Chlamydia trachomatis* in cell culture. The glycogen-containing inclusions stain dark brown when the cells are treated with an iodine solution (left). ©Photo Researchers, Inc/Alamy Stock Photo. When stained with fluorescein-labeled anti-*Chlamydia* antibody (right), the inclusions fluoresce brightly against a dark background when viewed with a fluorescence microscope (right). ©Josephine A. Morello

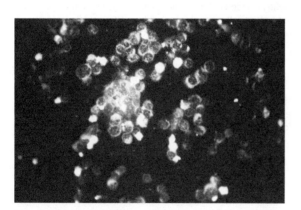

Plate 43 A direct immunofluorescence test performed on a nasal wash specimen. Respiratory syncytial virus is seen fluorescing brightly in the infected human epithelial cells.
©Scott Camazine/Alamy Stock Photo

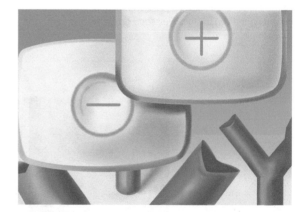

Plate 44 A direct enzyme immunoassay reveals the presence of respiratory syncytial virus in a nasal wash sample when the positive sign appears (right). Appearance of the negative sign instead (left) would indicate that no respiratory syncytial virus was detected in the sample.

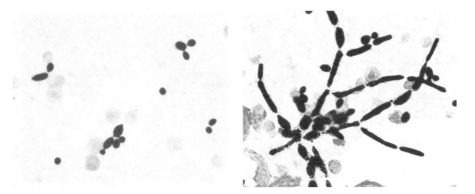

Plate 45 Gram stain of *Candida albicans* cells isolated from the blood culture of a patient. At left the yeast cells are budding, and at right, they have formed long, filamentous, irregularly staining hyphae. ©Josephine A. Morello

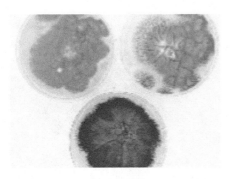

Plate 46 Colonies of three *Aspergillus* species. Some molds may be recognized by the color of their spores (conidia). Clockwise from left: *A. flavus* (yellow), *A. fumigatus* (smoky gray-green), *A. niger* (black). ©Josephine A. Morello

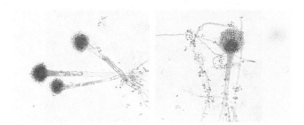

Plate 47 Lactophenol cotton blue coverslip preparation from a slide culture of *Aspergillus fumigatus*. At low power (×100) magnification (left), three spore-bearing structures can be seen. At higher power (×400) magnification (right), the characteristics that permit genus and species identification are clearly visualized. ©Josephine A. Morello

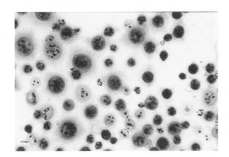
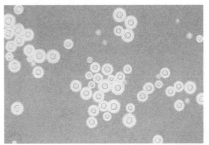

Plate 48 *Cryptococcus neoformans* in cerebrospinal fluid from a patient with AIDS. In the Gram-stained preparation at left, yeast cells are seen to stain irregularly. The orange-staining halo around some cells is the *Cryptococcus* capsule. In the India ink preparation at right, the cryptococcal yeast cells are surrounded by a capsule that is demarcated by the suspension of carbon particles in the India ink. Left: ©Josephine A. Morello; Right: Source: Dr. Leanor Haley, CDC

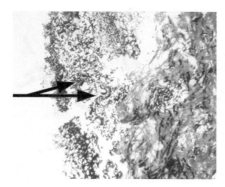
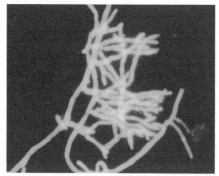

Plate 49   Left: Methenamine silver stain of sinus biopsy (×100). All of the black material represents the invading fungus, *Aspergillus flavus.* Two of the characteristic spore-bearing structures can be seen on the nasal cavity side (arrows). Right: A calcofluor stain of the biopsy material as seen under an ultraviolet light source (×400). ©Josephine A. Morello

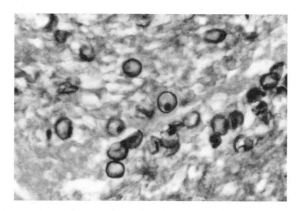

Plate 50   Wood's lamp exam showing fluorescence of infected hairs. ©Alamy Stock Photo

Plate 51   Methenamine silver stain of cyst form of *Pneumocystis jiroveci* from lung biopsy of a patient with AIDS. ©Centers for Disease Control and Prevention/Science Source

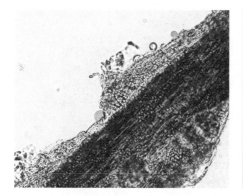
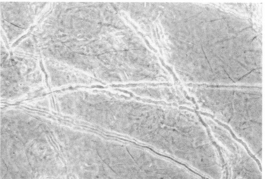

Plate 52   KOH preparations of a hair and skin scales from patients with tinea capitis (ringworm of the hair) and tinea corporis (ringworm of the body). On the left, many round, reproductive spores of the dermatophyte fungus are seen surrounding the hair; the filamentous hyphae invade the hair shaft. On the right, the filamentous hyphae are seen invading skin scales throughout the preparation. ©Josephine A. Morello

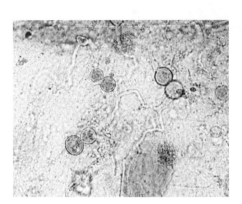

Plate 53 Lactophenol cotton blue preparation of *Blastomyces dermatitidis* growing in culture of sputum. The characteristic thick-walled, broad-based budding yeast cells are seen at the top right and bottom left of the preparation. ©Josephine A. Morello

Plate 54 Scanning electron micrograph of *Trichomonas vaginalis* showing the characteristic undulating membranes and flagella (×12,000). ©David M. Phillips/Science Source

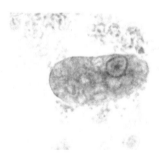

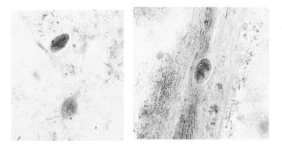

Plate 55 A trophozoite of *Entamoeba histolytica.* The characteristic circular nucleus at the top right portion of the ameba has dark chromatin evenly distributed around its edges and a centrally placed karyosome. ©Josephine A. Morello

Plate 56 Left panel: *Giardia lamblia* trophozoite (bottom) and cyst seen in stool specimen. At the right, the characteristic features of the cyst are revealed more clearly (×1,000). ©Josephine A. Morello

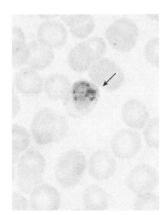

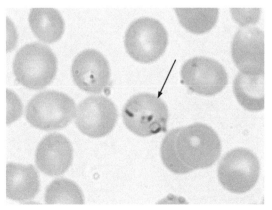

Plate 57 *Plasmodium* trophozoites in red blood cells (arrows). *P. vivax* at left causes the blood cell to enlarge and show characteristic stippling (Schüffner's dots). At right, three trophozoites (ring forms) of *P. falciparum* in a single red blood cell. Multiply infected cells are characteristic of this malarial species; the infected cell is not enlarged and no Schüffner's dots are present. Left: ©Dr. E.J. Bottone; Right: ©Josephine A. Morello

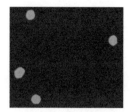

Plate 58 *Cryptosporidium* spp. are prevalent in animals but also infect humans, causing massive, watery diarrhea. Diagnosis can be made by finding oocysts in the patient's fecal specimen, either with a modified acid-fast stain (circular red objects, left) or with a specific fluorescent antibody reagent (circular green objects, right). Left: Source: Oregon State Health Department and CDC-DPDx; Right: ©Josephine A. Morello

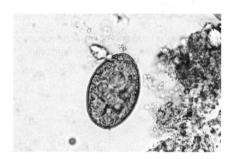

Plate 59 Characteristic egg of the fish tapeworm, *Diphyllobothrium latum,* ×400. Larvae in raw or undercooked fish mature to adulthood in the human intestinal tract and shed eggs that help provide the diagnosis. ©Josephine A. Morello

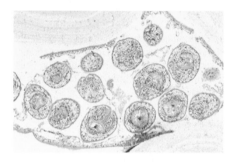

Plate 60 *Echinococcus granulosus* cyst from human liver. More than 12 larval forms of *Echinococcus* can be seen budding off from the thick-walled capsule.
©Josephine A. Morello

Table 19.1   Some Characteristics for Differentiating *Staphylococcus* Species

Characteristic	S. aureus	S. epidermidis	S. saprophyticus
Coagulase production	Positive	Negative	Negative
Pigment production (color of colonies)	Light golden yellow	White	Yellow or white
Hemolysis on blood agar	Usually positive	Negative	Negative
Mannitol fermentation	Positive	Negative	Usually negative
Novobiocin susceptibility	Susceptible	Susceptible	Resistant (inhibition zone ≤16 mm)
Catalase production*	Positive	Positive	Positive
Protein A production	Positive	Negative	Negative

*Note that the gram-positive streptococci, which may resemble staphylococci microscopically in Gram-stained smears made from agar media, are catalase negative.

environment. The problem is compounded by the fact that many hospital strains of *S. aureus* are resistant to most of the commonly used antimicrobial agents, especially methicillin. These strains are known as methicillin-resistant *S. aureus*, or MRSA. Methicillin-susceptible strains are referred to as MSSA. As described in Exercise 12, MRSA infections are occurring with increasing frequency in the community as well. MRSA infections that typically occur in hospitals are referred to as hospital-acquired MRSA (HA-MRSA) infections, whereas those that occur in the community are called community-acquired MRSA, or CA-MRSA infections.

To select the correct antimicrobial therapy and institute appropriate infection-control measures, when indicated, it is important that laboratories quickly differentiate MRSA from MSSA. Several methods have been developed solely for this purpose. These include the use of a commercially prepared selective and differential medium that distinguishes MRSA from MSSA isolates on the basis of the color of the colony growing on this medium. In another test, a latex agglutination assay is used to screen for the presence of the protein PBP 2a, which is produced by the *mecA* gene in *S. aureus*. The *mecA* gene is responsible for methicillin resistance. In another molecular assay, the polymerase chain reaction (PCR) is used to specifically detect the *mecA* gene in *S. aureus* strains present in nasal swabs, blood cultures, or wound specimens.

All personnel involved in patient care should be knowledgeable of transmission routes and follow strict procedures designed to prevent nosocomial infection and spread of these strains in the population. In the experiments that follow, you will be seeing and handling staphylococcal cultures. Use your knowledge of aseptic technique and make certain that you do not carry staphylococci out of the laboratory as new additions to the flora of your hands or clothes. Keep your hands scrupulously clean. If you have any minor cuts or scratches or other injury to your hands, they should be protected with gloves. While in the laboratory, keep your hands and implements with which you are working away from your mouth and face.

Purpose	To isolate and identify staphylococci
Materials	Blood agar plate (BAP)
	Mannitol salt agar (MSA) plate
	Tubed plasma (0.5-ml aliquots)
	Novobiocin disks (5 μg)
	Sterile 1.0-ml pipettes
	Pipette bulb or other aspiration device
	Latex agglutination kit for *Staphylococcus aureus*
	24-hour broth cultures of *Staphylococcus epidermidis, Staphylococcus aureus, Staphylococcus saprophyticus,* and *Escherichia coli*
	Wire or disposable inoculating loop
	Marking pen or pencil

Procedures

1. With your marking pencil, divide the bottom of the BAP and the MSA plate into four segments each. Label each segment with the name of one of the four organisms in the broth cultures.
2. Using the grown broth cultures, inoculate one section of each appropriately labeled plate with *S. aureus,* one with *S. epidermidis,* one with *S. saprophyticus,* and one with *E. coli.* Streak each section carefully, remaining within the assigned space.
3. With heated and cooled forceps, pick up a novobiocin disk, place it in the center of one of the streaked areas of the BAP, and press it gently onto the agar with the forcep tips.
4. Repeat step 3 for the remaining three organisms on the streaked BAP.
5. Place the plates in the 35°C incubator for 24 hours.
6. Perform a coagulase test on each of the three *Staphylococcus* broth cultures as follows:
   a. Using a sterile pipette, measure 0.1 ml of the *S. epidermidis* broth culture with the aspiration device. Transfer this inoculum to a tube of plasma. Discard the pipette in disinfectant or a biohazard container. Label the tube.
   b. Inoculate a second and third tube of plasma with 0.1 ml of the *S. aureus* and *S. saprophyticus* broth cultures, respectively, as in step 6a.
   c. Place all inoculated plasma tubes in the 35°C incubator. After 30 minutes, remove and examine them (close the incubator door while you read them). Hold the tubes in a semihorizontal position to see whether the plasma in the tube is beginning to clot into a solid mass. If so, make a record of the tube showing coagulase activity. Return unclotted tubes to the incubator.
   d. Repeat procedure 6c every 30 minutes for 4 hours, if necessary.
7. After 24 hours of incubation of the plate cultures prepared in procedures 1 and 2, examine and record colonial morphology. Make Gram stains of each culture on the BAP and record microscopic morphology. Measure and record the diameter of the zone of inhibition around the novobiocin disks. A zone size greater than 16 mm in diameter is considered susceptible.
8. Following the instructor's directions, place one drop of the latex agglutination reagent onto each of two circles on the card provided. With the special stick contained in the kit or a sterile inoculating loop, pick up several colonies of *S. aureus* from the blood agar plate you inoculated at the previous laboratory session. Emulsify the colonies in the latex reagent, being careful not to scratch the card. Repeat this procedure with colonies of *S. epidermidis. Do not use colonies from the mannitol agar plate as these are difficult to emulsify.*
9. Rotate the card gently for 20 seconds, observing the circles for a clearly visible clumping of the latex particles and a clearing of the milky background (see fig. 19.1). This reaction signifies a positive test. Record the results in the table, then dispose of the reaction card in a biohazard container or the disinfectant provided.

Results

1. Table of plate culture results.

Name of Organism	Colonial Morphology on BAP	Gram-Stain Reaction and Microscopic Morphology on BAP	Appearance on MSA	Novobiocin Zone Diameter
Staphylococcus epidermidis				
Staphylococcus aureus				
Staphylococcus saprophyticus				
Escherichia coli				

2. Results of identification tests.

Name of Organism	Coagulase (+ or −)	Time Required to Clot Plasma	Appearance of Plasma After 4 Hours	MSA Reaction for Mannitol Fermentation* (+ or −)	Novobiocin* (S or R)	Latex Agglutination (+ or −)
Staphylococcus epidermidis						
Staphylococcus aureus						
Staphylococcus saprophyticus						

*Your interpretation of results from MSA and novobiocin plates.

EXPERIMENT 19.2 Staphylococci in the Normal Flora

Purpose	To isolate and identify staphylococci in cultures of the nose and hands
Materials	Blood agar plates (BAP)
	Mannitol salt agar (MSA) plates
	Sterile swabs
	Dropping bottle containing hydrogen peroxide
	Tubed plasma (0.5-ml aliquots)
	Latex agglutination kit for *Staphylococcus aureus*
	Marking pen or pencil

Procedures

1. Take a culture of your own nose by swabbing the membrane of one of your anterior nares with a sterile swab.
2. Inoculate the nasal swab across the top quarter of a blood agar and a mannitol salt agar plate. Streak across the remainder of each plate for isolation of colonies. Discard the swab in disinfectant or a biohazard container.
3. Take a culture from the palm of your left hand by swabbing across it. Inoculate a blood agar and a mannitol salt agar plate and streak for isolation of colonies.
4. Sterilize your inoculating loop and moisten it in sterile saline. Run the moistened loop under one of your fingernails, picking up some debris if possible. Inoculate a blood agar and a mannitol salt agar plate and streak out.
5. Be certain all plates are labeled appropriately. Incubate all plates at 35°C for 24 hours.
6. Examine the plates and make Gram stains of different colony types on both the blood and mannitol plates. Perform a catalase test on the different colony types on both blood and mannitol plates by smearing a small amount of colony growth onto a slide and with a capillary pipette, placing one drop of hydrogen peroxide onto each smear. *Be careful not to dig into the blood agar medium or a false-positive result may be obtained.* Observe for bubble formation (refer to fig. 15.1). Perform a coagulase test on a colony of mannitol-positive staphylococci, if present, using your loop to pick it from an MSA plate and emulsify it directly in 0.5 ml of plasma (continue as in Experiment 19.1, steps 6c and d). Perform a rapid latex agglutination test on any beta-hemolytic colonies growing on BAP that show gram-positive cocci in clusters on Gram stain.
7. Record your observations in the following table.

Results

Body Site	Results from Blood Agar Plate			Latex Agglutination (+ or −) (Beta-hemolytic on BAP)	Results from Mannitol Salt Agar (MSA) Plate				Tentative Identification*
	Colony Morphology on Blood Agar Plate	Gram-Stain Reaction and Microscopic Morphology	Cata-lase (+ or −)		Colony Appearance on Mannitol Salt Agar	Microscopic Morphology	Catalase (+ or −)	Coagulase (+ or −) (if Mannitol +)	

*In the last column, indicate the tentative identification you would make of each colony described in the table.

CASE STUDY

Community-Acquired Methicillin-Resistant *Staphylococcus*

The following are two cases reported by the Centers for Disease Control and Prevention. They illustrate the emerging problem of community-acquired methicillin-resistant *Staphylococcus aureus,* as well as clinical and laboratory features of staphylococcal infection.

Case 1.

A 7-year-old girl was admitted to the hospital with a temperature of 103°F (39.5°C) and complaining of right groin pain. She was diagnosed with an infected right hip joint, which was drained surgically. She was treated with a cephalosporin-class antimicrobial agent, but on the third day, her antimicrobial therapy was changed to vancomycin when blood and joint-fluid cultures grew MRSA. The isolate was resistant to cephalosporins and methicillin, but was otherwise multidrug susceptible. After another hip drainage procedure, she developed acute respiratory distress syndrome, pneumonia, and an empyema (lung abscess) that required chest tube drainage. She died from a pulmonary hemorrhage after 5 weeks of hospitalization.

Case 2.

A 16-month-old girl was brought to a local hospital with fever, hemoptysis (bloody sputum), and respiratory distress. The day before admission, she had a productive cough and a 2-cm papule on her lower lip. A chest X-ray showed infiltrates in the left lower lobe of her lung and a pleural effusion (fluid in the pleural space). She was treated with a cephalosporin-type antimicrobial and a methicillin-type antimicrobial. Shortly after arriving at the hospital, her blood pressure dropped, and she was intubated and treated with vancomycin. Despite supportive efforts, she died after 7 days from progressive cerebral edema (brain swelling) and multiorgan failure. The patient's blood, sputum, and pleural fluid grew MRSA that was multidrug susceptible.

1. What was the likely source of the original *Staphylococcus* infection in each patient?
2. What two tests could the laboratory have used to identify these isolates as *Staphylococcus aureus?*
3. What antibiotic susceptibility pattern separates these isolates from MRSA isolated from hospitalized patients?
4. What precautions did hospital personnel need to take to prevent transmission of these strains to other hospitalized patients?

Source: Adapted from C. Hunt et al. 1999. Four pediatric deaths from community-acquired methicillin-resistant *Staphylococcus aureus*—Minnesota and North Dakota, 1997–1999. *MMWR Weekly* 48:707–710.

Questions

1. Differentiate the microscopic morphology of staphylococci and streptococci as seen by Gram stain.

2. What are the two types of staphylococcal coagulase?

3. What is protein A? Describe one method for detecting it.

4. What properties of *S. aureus* distinguish it from *S. epidermidis* and *S. saprophyticus?*

5. How is *S. saprophyticus* distinguished from *S. epidermidis?*

6. From what specimen type would *S. saprophyticus* most likely be isolated?

7. What is a nosocomial infection? Who acquires it? Why?

8. Why are staphylococcal infections frequent among hospital patients?

9. Discuss the role played by *S. aureus* in human infectious diseases.

Streptococci, Pneumococci, and Enterococci

Learning Objectives

After completing this exercise, students should be able to:

1. Describe the types of streptococcal hemolysis.
2. Discuss the best culture media and atmospheric and incubation conditions for cultivating streptococci.
3. Compare normal flora of the upper and lower respiratory tracts.
4. List diseases caused by the beta-hemolytic streptococci.
5. Explain how *Streptococcus pneumoniae* is distinguished from other streptococci that have the same hemolytic properties.

The mucous membranes of the upper respiratory tract that are exposed to air and food (nose, throat, mouth) normally display a variety of aerobic and anaerobic bacterial species: gram-positive cocci (*Streptococcus, Staphylococcus,* and *Peptostreptococcus* species); gram-negative cocci (*Neisseria, Moraxella,* and *Veillonella* species); gram-positive bacilli (*Corynebacterium, Propionibacterium,* and *Lactobacillus* species); gram-negative bacilli (*Haemophilus, Prevotella,* and *Bacteroides* species); and, sometimes, yeasts (*Candida* species).

These flora vary somewhat in different areas of the upper respiratory tract. In the nasal membranes, staphylococci predominate. The throat (pharyngeal membranes) has the widest variety of microbial species, and the sinus membranes have few, if any, organisms. The lower parts of the respiratory tract (trachea, bronchi, alveoli) are not readily colonized by microorganisms, because the ciliated epithelium of the upper respiratory tract membranes, together with mucous secretions, trap and move them upward and outward.

Usually the normal flora commensals of the upper respiratory tract prevent entry and colonization by transient microorganisms. Some of these intruders might be pathogenic and subsequently invade lower respiratory tissues to produce disease. Infection may occur when the healthful conditions maintained by the normal flora are disturbed, for example, by changes in the host's immune status, administration of antimicrobial agents to which the commensals are susceptible, and exposure to large numbers of virulent pathogens.

The Streptococci

The genus *Streptococcus* contains gram-positive cocci that characteristically are arranged in chains (see **colorplate 2**). A number of species of streptococci are normally found among the normal flora of human skin and mucous membranes, particularly those of the upper respiratory tract. Certain species are more commonly associated with human infectious diseases than others.

Many streptococci have fastidious growth requirements including a requirement for blood-enriched media. Most grow well in air but also grow in the absence of oxygen (i.e., they are *facultative anaerobes*), some prefer reduced oxygen tension and increased CO_2 (*microaerophilic*), and some grow only in the absence of oxygen (*anaerobic*). The anaerobic streptococci are now placed in the genus *Peptostreptococcus* (see Exercise 28). An incubation temperature of 35°C is optimal for growth of most streptococci.

A number of streptococcal species produce substances that destroy red blood cells; that is, they cause *lysis* of the red cell wall with release of hemoglobin. Such substances are referred to as *hemolysins*. The activity of streptococcal hemolysins (also known as *streptolysins*) can be readily observed when the organisms are growing on a blood agar plate.

Different streptococci produce different effects on the red blood cells in blood agar. Those that produce *incomplete* hemolysis and only partial destruction of the cells around colonies are called *alpha-hemolytic streptococci.* Characteristically, this type of hemolysis is seen as a distinct greening of the agar in the hemolytic zone, and thus this group of streptococci has also been referred to as the *viridans* group (from the Latin word for *green*). *Streptococcus pneumoniae,* also known as the pneumococcus, is a potential pathogen that also produces alpha hemolysis, as seen in **colorplate 31**. It may be found as normal flora in the throats of some persons, but is considered the cause of disease when isolated in large numbers from the sputum and blood of patients with clinical signs of pneumonia. Pneumococci will be studied in more detail in Experiment 20.3.

Species whose hemolysins cause *complete* destruction of red cells in the agar zones surrounding their colonies are said to be *beta-hemolytic.* When growing on blood agar, beta-hemolytic streptococci are small opaque or semitranslucent colonies surrounded by clear zones in an otherwise red-colored, opaque medium (see **colorplate 11**). One of the two streptococcal hemolysins involved in this reaction is inhibited by oxygen. Its effect is seen best around subsurface colonies or when culture plates are incubated anaerobically. Some strains of staphylococci, *Escherichia coli,* and other bacteria also may show beta-hemolysis.

Some species of streptococci do not produce hemolysins. Therefore, when their colonies grow on blood agar, no change is seen in the red blood cells around them. These species are referred to as *nonhemolytic* streptococci, although formerly, they were called *gamma* streptococci. **Colorplate 28** shows the appearance of alpha-, beta-, and nonhemolytic streptococci growing as subsurface colonies on blood agar plates. Alpha-, beta-, and gamma-hemolytic streptococci are often referred to by their Greek letter designations, α, β, and γ, respectively.

In the clinical microbiology laboratory, observation of hemolysis is an important first step in differentiating among streptococcal species. Most alpha-hemolytic streptococci are members of the normal throat flora and do not need to be identified further when they are isolated from respiratory cultures. As stated earlier, the exception is the pneumococcus, which is identified when isolated in culture from sputum of patients with clinical pneumonia. Also, in persons with heart valve abnormalities, alpha-hemolytic streptococci may be deposited on the valves, usually at the time of dental work when they have the opportunity to enter and circulate through the bloodstream. The inflammatory process that develops on the valve is referred to as *endocarditis,* and if untreated, is a life-threatening infection. In patients with endocarditis, isolation of viridans streptococci from multiple blood cultures is considered diagnostic of endocarditis.

When beta-hemolytic streptococci are found in throat cultures, the laboratory must proceed with further testing to determine the antigenic group. This is done by extracting the carbohydrate antigen from the streptococcal cell wall and reacting it with specific antibodies in a latex agglutination test or enzyme immunoassay (see Exercise 16). The most important streptococcal group is group A, which is responsible for streptococcal pharyngitis ("strep throat") and a variety of other serious skin and deep tissue infections (see table 20.1). The species name given to group A streptococci is *pyogenes* (pus producing, a characteristic of the infection produced). Certain toxins and extracellular products of *S. pyogenes* are responsible for scarlet fever, a toxic shock syndrome similar to that produced by *Staphylococcus aureus,* and the flesh-eating disease (necrotizing fasciitis). In addition, rheumatic fever is caused by immunologic reactions between streptococcal antibodies and heart tissue antigens, and glomerulonephritis occurs as a result of antigen-antibody complex deposition in the kidney glomeruli. These diseases are referred to as *sequelae* of group A streptococcal infection.

At one time, group B streptococci (*Streptococcus agalactiae*) were considered primarily animal pathogens. Now, they are known to colonize the human female vaginal tract. In some colonized pregnant women, the organism causes infection of the endometrium following delivery, and more seriously, may produce sepsis and meningitis in the newborn child. Because of these severe infections, pregnant women are routinely screened for vaginal carriage of the group B *Streptococcus* a few weeks before term, and treated with antimicrobial agents during labor if they are colonized. Group B streptococci also are found in men and nonpregnant women, in whom they may cause urinary tract and wound infections, osteomyelitis, and pneumonia.

Other beta-hemolytic streptococci are placed in groups C through V, but most do not cause disease. Groups C, F, and G may cause mild pharyngitis but do not have the serious effects that groups A and B do.

Streptococcal-like bacteria with group D antigen were at first classified in the genus *Streptococcus,* but studies have revealed that they differ in many biological respects. Therefore, they have now been placed in their own genus, *Enterococcus.*

Diagnostic Microbiology in Action

Figure 20.1 A positive latex agglutination reaction for group A streptococci. The left-hand well shows a granular precipitate. In this well, the group A carbohydrate antigen has combined with latex beads coated with antibody against this specific antigen. The well on the right (group B antigen) remains negative, showing only the fine suspension of nonagglutinated latex particles. This antigen does not react with the anti–group A antibodies on the latex particles.
©Verna Morton

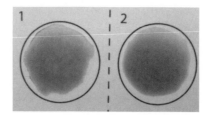

Identification of Streptococci

In smears from patient material, the microbiologist presumptively identifies gram-positive cocci in chains as streptococci (although usually, the Gram-stain reaction and morphology only are reported to the physician). When culture growth is available, the type of hemolysis produced by colonies on blood agar plates leads to the next step(s) in identification. Alpha-hemolytic colonies from upper respiratory specimens are not identified further because they are considered normal flora. Beta-hemolytic colonies must be identified further, usually by serological methods, to determine whether or not they are group A (any specimen type) or group B (genital specimens from pregnant women).

Serological testing is most commonly done by the latex agglutination method (see fig. 16.2) and is the definitive method for grouping beta-hemolytic streptococci. An alternative test is based on the ability of *S. pyogenes*, but not other beta-hemolytic streptococci, to hydrolyze the reagent PYR (L-pyrrolidonyl-beta-naphthylamide) to a bright red color (see **colorplate 33**). Members of the *Enterococcus* genus also hydrolyze PYR. Historically, *S. pyogenes* was identified presumptively by its susceptibility to the drug bacitracin. In this test, the unknown organism was streaked onto a portion of a blood agar plate and a bacitracin disk was placed on top of the inoculum. After incubation, if the growth of beta-hemolytic colonies was inhibited around the disk, the organism was identified presumptively as a group A streptococcus (see **colorplate 29**). Group B streptococci produce a substance called the CAMP factor, which enhances the effect of a beta-hemolysin produced by some strains of *Staphylococcus aureus*. In a positive CAMP test, a bright arrow of hemolysis is seen when a group B streptococcus is streaked at right angles to the appropriate *S. aureus* strain (see **colorplate 30**). Both the bacitracin and CAMP tests are seldom performed at present because false-positive and -negative reactions occur, and more-specific, rapid serologic tests are available.

In addition to the rapid agglutination test used for identifying group A streptococci from culture (fig. 20.1), a large number of rapid immunoassays are available for detecting the group A antigen directly from throat swabs (see figures 16.3 and 16.4). These tests are usually performed in clinics and physicians' offices—that is, at the "point of care," because they are rapid (10 to 30 minutes) and do not require culture expertise. However, for a positive test, a large number of organisms is needed on the swab. Because of this, false-negative immunoassay results may occur if a low number of *S. pyogenes* organisms is present in the throat swab specimen. Thus, when negative results are obtained for patients with clinical evidence of pharyngitis, a throat swab for "strep" culture should always be sent to the clinical microbiology laboratory.

Gene amplification assays, such as PCR (see Exercise 17), are also being used increasingly for the detection of *S. pyogenes* directly in pharyngeal specimens. Like immunoassays, the PCR test can be completed in less than 1 hour. In addition, because the PCR test is considerably more sensitive than immunoassays and culture, a negative PCR result for *S. pyogenes* is considered final, and a subsequent throat culture is not needed.

Table 20.1 summarizes the characteristics of streptococci, pneumococci, and enterococci.

Table 20.1 Classification and Identification of Streptococci, Pneumococci, and Enterococci

Experiment No	Serological Group*	Organism Name	Type of Hemolysis	Cellular Products	Identification	Clinical Diseases
20.1	A	S. pyogenes	Beta	Group A carbohydrate Streptolysins Scarlet fever (pyrogenic) toxin Deoxyribonuclease Toxic shock syndrome toxin	Susceptible to bacitracin (A disk) Positive PYR reaction Serological typing PCR test	Pharyngitis-tonsillitis Skin infections Scarlet fever Postpartum endometritis Necrotizing fasciitis Rheumatic fever Toxic shock syndrome Glomerulonephritis
20.2	B	S. agalactiae	Beta	Group B carbohydrate	Positive CAMP test Serological typing	Neonatal sepsis and meningitis Postpartum endometritis Osteomyelitis Wound infection Urinary tract infection Pneumonia
	C	S. equisimilis	Beta	Group C carbohydrate	Serological typing	Pharyngitis
20.4	D	Enterococcus spp.	Alpha, beta, or nonhemolytic	Group D carbohydrate	Growth in 6.5% salt broth Growth and blackening on bile-esculin medium Positive PYR reaction Serological typing	Endocarditis Urinary tract infection Wound infection
20.3	Not grouped	Viridans streptococci	Alpha		Hemolytic reaction Bile insoluble Resistant to optochin (P disk) Biochemicals if necessary	Endocarditis
20.3	Types 1–80+	S. pneumoniae	Alpha	Capsular carbohydrate	Susceptible to optochin (P disk) Bile soluble Serological (capsular) typing	Pneumonia Bacteremia Meningitis Endocarditis

*Note: The basis for grouping beta-hemolytic streptococci is a carbohydrate found in the cell wall; pneumococci are typed according to the carbohydrate capsule exterior to the cell wall.

EXPERIMENT 20.1   Identification of Group A Streptococci (*Streptococcus pyogenes*)

In this experiment, we will study some simple methods for isolating streptococci from clinical specimens and for confirmatory identification of beta-hemolytic strains as group A.

Purpose	To isolate and identify streptococci in culture
Materials	Sheep blood agar plate (BAP)
	Simulated throat swab specimen from a 2-year-old child with acute tonsillitis
	Kit for detection of group A streptococcal antigen directly from throat swab
	Demonstration blood agar plate showing alpha-hemolytic, beta-hemolytic, and nonhemolytic strains of streptococci
	Solution with extracted antigen of beta-hemolytic *Streptococcus* (prepared by instructor)
	Latex test kit for serological typing
	Wire inoculating loop
	Marking pen or pencil

Procedures

1. Inoculate and streak a labeled blood agar plate with the simulated clinical specimen. Make a few stabs in the agar at the area of heaviest inoculum. Try not to stab to the bottom of the agar medium layer.
2. Incubate the plate at 35°C for 24 hours.
3. Test the simulated patient throat swab specimen with the test kit for detecting group A streptococcal antigen *directly* from throat swabs. Follow the manufacturer's instructions carefully.
4. Record your observations under Results (no. 1).
5. Examine the demonstration plate (but *do not open* it without supervision).
6. Following the manufacturer's directions, use the latex typing kit to identify serologically the beta-hemolytic isolate. Mix the antigen extract prepared by the instructor with a drop of each of the group A and group B latex reagents.
7. Observe both suspensions for evidence of agglutination. See fig. 20.1.
8. Record your observations under Results (no. 3).

Results

1. Group A antigen test performed directly from the simulated throat swab (+ or −)?_____
2. From the demonstration plate showing different types of hemolysis, describe your observations of

alpha-hemolytic streptococci _____

_____

beta-hemolytic streptococci _____

_____

nonhemolytic streptococci _____

_____

3. With which latex reagent in the serological typing kit did you obtain a positive result? Group _____

4. Examine the incubated BAP of the "patient's" throat culture. Describe the results in the following table, after making Gram stains.

Morphology of Individual Colonies	Type of Hemolysis Displayed	Gram-Stain Reaction and Microscopic Cell Morphology

5. Describe any differences in intensity of hemolysis around colonies growing on the agar surface and those pushed below the surface where you stabbed into the agar.
6. How would you report the culture results to the physician?

CASE STUDY

A Case of Pharyngitis

A 6-year-old male awoke complaining of throat pain and difficulty swallowing. His temperature was 102°F (39°C) and his mother could feel "swollen glands" in his neck. When she looked into his mouth, she saw whitish patches over his tonsils. She immediately brought him to his pediatrician, who sent a throat swab for direct antigen testing. Upon receiving the test results, the physician prescribed penicillin for 10 days. The child felt better within 48 hours, returned to school, and recovered uneventfully.

1. If the direct antigen test result was negative, what is the next best course of action to take?
2. If the swab was cultured for streptococci, what culture medium should the laboratory have used?
3. On this medium, what would be the appearance of colonies of the organism causing the child's pharyngitis?
4. How would the identification of the bacterial agent be confirmed?
5. Why was penicillin prescribed instead of a newer antimicrobial agent?

EXPERIMENT 20.2   The CAMP Test for Group B Streptococci (*Streptococcus agalactiae*)

Group B streptococci can be distinguished from other beta-hemolytic streptococci by their production of a substance called the CAMP factor. This term is an acronym for the names of the investigators who first described the factor: *C*hristie, *A*tkins, and *M*unch-*P*etersen. The substance is a peptide that acts together with the beta-hemolysin produced by some strains of *Staphylococcus aureus,* enhancing the effect of the latter on a sheep blood agar plate. This effect is sometimes referred to as *synergistic* hemolysis (see **colorplate 30**). A serological test will also be performed.

Purpose	To differentiate group B from group A streptococci by the effect of the group B CAMP factor and by a serological method
Materials	Demonstration sheep blood agar plate, streaked at separate points with *Staphylococcus aureus,* group B streptococci, and group A streptococci
	Solution with extracted antigen of beta-hemolytic *Streptococcus* (prepared by instructor)
	Latex test kit for serological typing
	Wire inoculating loop
	Marking pen or pencil

Procedures (steps 1–4 to be performed by instructor)

1. With an inoculating loop, streak a strain of *S. aureus* down the center of a blood agar plate labeled CAMP test. (ATCC 25923 or other strain known to produce beta-hemolysin is used; 5% sheep blood agar is needed.)
2. On one side of the plate, inoculate a strain of group B *Streptococcus* by making a streak at a 90° angle, starting 5 mm away from the *S. aureus* and extending outward to the edge of the agar (see diagram in Results).
3. On the other side of the plate, inoculate a strain of group A *Streptococcus,* again at a 90° angle from the *S. aureus,* as in step 2. This streak should not be directly opposite the group B inoculum (see diagram in Results).
4. Incubate the plate aerobically at 35°C for 18 to 24 hours.
5. The student should confirm the isolate's identity by the serological test. Using the extracted antigen solution, follow the procedures in steps 6 and 7 of Experiment 20.1.

Results

1. Observe the area of hemolysis surrounding the *S. aureus* streak. At the point adjacent to the streak of group B streptococci, you should see an arrowhead-shaped area of increased hemolysis indicating production of the CAMP factor (review **colorplate 30**). There should be no change in the hemolytic zone adjacent to the streak of group A streptococci, most strains of which do not produce the CAMP factor.

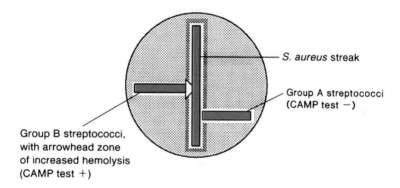

2. Although most group A streptococci give a negative CAMP test, some have been reported to be positive, especially when the test plate has been incubated anaerobically rather than aerobically. Most laboratories now use a streptococcal typing kit to confirm the identification of group B streptococci, because the test is rapid and confirmatory and can be performed with one or two colonies. In contrast, demonstration of the CAMP reaction requires overnight incubation.
3. CAMP reaction (+ or −) _____. Latex agglutination (+ or −) A _____ B _____.

EXPERIMENT **20.3**   **Identification of Pneumococci (*Streptococcus pneumoniae*)**

Pneumococci are among the most important agents of bacterial pneumonia. Other microorganisms such as staphylococci (Exercise 19), *Haemophilus influenzae* (Experiment 21.1), and *Klebsiella pneumoniae* (Exercise 23) may also be associated with serious pulmonary disease. Bacterial agents of pneumonia cause an acute inflammation of the bronchial and/or alveolar membranes. When the alveoli are involved, fluid accumulation may rupture the thin alveolar membranes. The fluid results from hemorrhage of alveolar capillaries and infiltration of inflammatory exudate containing many white blood cells (pus). Laboratory diagnosis is often made by isolating the causative agent from *sputum,* a secretion from the lower respiratory tract, sent for culture. However, because sputum specimens pass through the oropharynx as they are expectorated, contaminating members of the normal throat flora may interfere with culture results by overgrowing the pathogen. The causative organism is often found in the bloodstream during early stages of infection, and therefore, patient blood should also be cultured when pneumonia is suspected. In some patients, the organisms spread from the bloodstream to the central nervous system to cause meningitis, and pneumococci can then be isolated from the patient's cerebrospinal fluid (CSF) as well. In recent years, the incidence of serious, life-threatening pneumococcal infections has been significantly reduced by the use of vaccines that protect against infections caused by the more common serotypes of *S. pneumoniae*. Two examples of these vaccines are Pneumovax, which protects against 23 types of pneumococci, and Prevnar 13, which protects against 13 types.

Pneumococci are classified in the genus *Streptococcus* as the species *pneumoniae*. They are gram-positive, lancet-shaped cocci that characteristically appear in pairs (diplococci) or in short chains (see **colorplate 3**). Like other streptococci, they are fastidious microorganisms and require blood-enriched media and microaerophilic conditions for primary isolation. They are alpha-hemolytic and usually produce greening of blood agar around their colonies. *Streptococcus pneumoniae* can be distinguished from other alpha-hemolytic streptococci because it is lysed by bile salts and other surface active substances, including one known as optochin (see **colorplate 31**).

Another distinctive feature of pneumococci is that they possess a capsule composed of a viscous polysaccharide. This slimy capsule protects them from destruction by phagocytes that gather at sites of infection to ingest them. In the laboratory, the pneumococcal capsules are not readily demonstrated by usual staining techniques, but they can be made visible under the microscope by a serological technique known as the quellung reaction. *Quellung* is the German word for "swelling" and describes the microscopic appearance of pneumococcal or other bacterial capsules after their polysaccharide antigen has combined with a specific antibody present in a test serum from an immunized animal. As a result of this specific combination, the capsule appears to swell and its outlines become clearly demarcated (see **colorplate 10**).

The capsular antigen can also be detected with antibody-coated latex reagents. Colonies of suspected pneumococci growing on blood agar plates may be tested, or, depending on the disease severity, the soluble capsular antigen may be present in the patient's CSF, blood, and urine (the antigen, but not necessarily the organisms, is excreted from the body by the kidneys). Regardless of the results of direct antigen detection tests, cultures of sputum, blood, and CSF (in patients with signs and symptoms of meningitis) should always be performed. In some instances, the antigen concentration in body fluids is too low to be detected, but cultures are positive. In contrast to the beta-hemolytic streptococci, which remain susceptible to penicillin, the emergence of pneumococcal resistance to penicillin and other beta-lactam antibiotics dictates that isolates of *S. pneumoniae* be tested by CLSI-approved susceptibility testing methods.

Purpose	To identify pneumococci in culture
Materials	Dropping bottle containing 10% sodium desoxycholate or sodium taurocholate (bile solution) Optochin disks Forceps Blood agar plate Candle jar Blood agar plate cultures of pneumococci and other alpha-hemolytic streptococci Marking pen or pencil

Procedures

1. Examine the blood agar plate cultures of pneumococci and of alpha-hemolytic streptococci and note any differences in colonial morphology. Make a Gram stain of each organism.
2. Mark a blood agar plate with your marking pencil to divide it in half. Label one half "pneumo" and the other half "alpha-strep." Streak the appropriate sides heavily with a loopful of the pneumococcus and alpha-hemolytic streptococcus cultures.
3. Flame or heat your forceps lightly and use them to take up an optochin disk. Place the disk in the center of the area on the blood plate you streaked with pneumococci. Reheat the forceps and place another disk in the middle of the section streaked with alpha-hemolytic streptococci. Press each disk down lightly on the agar with the tip of the forceps, to make certain it is in contact and will not fall off when the plate is inverted (do not press it *through* the agar). Reheat the forceps.

Note: Optochin is the commercial name for ethylhydrocupreine hydrochloride, a surface reactant impregnated in the disk. Its effect on pneumococcal cell surfaces is similar to that of bile. The disk is often called a "P-disk" because it is used to distinguish susceptible *p*neumococci from other streptococci that are not lysed by surface reactants.

4. Invert the plate and place it in a candle jar. Light the candle, replace the lid of the jar (tightly), and wait for the candle flame to burn out. Place the jar in the 35°C incubator for 24 hours. Any wide-mouthed, screw-cap jar can serve as a candle jar. The candle burning in the closed jar uses up some of the oxygen and increases the carbon dioxide level. At a certain point, the oxygen is not sufficient for the candle to continue burning, and the flame is extinguished. The atmosphere remaining within the jar contains the increased carbon dioxide tension and the reduced oxygen tension preferred by many bacterial species, such as pneumococci, when they are first removed from the body and cultured on artificial medium (fig. 20.2). In many clinical laboratories, the plates are incubated in a special CO_2 incubator. Gas flowing into the incubator from a CO_2 cylinder maintains a constant level of 5 to 7% CO_2.

5. After overnight incubation, remove the plate from the CO_2 incubator or candle jar. Observe any macroscopic differences in the colonies of the *S. pneumoniae* and the alpha-hemolytic streptococcus isolates.

6. Add a drop of the bile solution to each side of the incubated plate and incubate the plate with its lid slightly ajar for 30 minutes at 35°C. Record your results in the Results section under no. 2. Bile-soluble colonies disintegrate and are no longer visible as discrete colonies, a result indicating the isolate is *S. pneumoniae*. Bile-insoluble colonies remain intact, indicating that the isolate is not *S. pneumoniae*.

Figure 20.2 A closed candle jar containing Petri plate cultures. The candle flame has gone out because the remaining oxygen is not sufficient to keep the flame lit. The jar now contains increased carbon dioxide and decreased oxygen, an atmosphere (microaerophilic) preferred by many bacteria. ©*Josephine A. Morello*

Results

1. Record your observations of the colonial and microscopic morphology of pneumococci and other alpha-hemolytic streptococci in the table following step 3 below.

2. Record results of the bile solubility test in the table. Describe here the appearance of each colony type at the end of the 30-minute test.

   Pneumococcus colonies with bile _____

   without bile _____

   Alpha-hemolytic *Streptococcus* colonies with bile _____

   without bile _____

3. Record results of the optochin disk test in the following table. Diagram the appearance of the growth on the plate with the disks, indicating your interpretation.

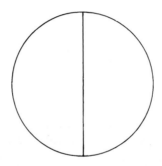

Name of Organism	Colonial Morphology (and Hemolysis)	Gram-Stain Reaction and Microscopic Morphology	Bile Solubility (+ or −)	Optochin Susceptibility (+ or −)
Streptococcus pneumoniae				
Alpha-hemolytic *Streptococcus*				

CASE STUDY

A Case of Pneumonia with Meningitis

A 78-year-old woman developed rapid onset of fever and a shaking chill. She had difficulty breathing and produced thick (mucopurulent), rust-colored sputum. Shortly thereafter, she developed a headache, lethargy, and a stiff neck and lapsed into a coma. She was admitted to the hospital where blood, sputum, and cerebrospinal fluid specimens were obtained for culture. The laboratory shortly reported that Gram-stained smears of the sputum and cerebrospinal fluid revealed gram-positive, lancet-shaped diplococci. The patient was treated with intravenous antimicrobial agents, woke up from the coma, and recovered without incident.

1. What was the likely identification of the patient's isolate?
2. What type of hemolysis would it produce on blood agar cultures?
3. What is the major virulence factor of this organism?
4. Is there a mechanism available to help prevent this type of disease? Describe.

EXPERIMENT 20.4   Identification of Enterococci

Enterococci are gram-positive cocci that form chains in culture and that were once classified in the genus *Streptococcus*. Because they differ in several characteristics, including the composition of their genetic material, they are now classified in a separate genus, *Enterococcus*. As the name implies, enterococci are found primarily in the intestinal tract, although they may be found in the upper respiratory tracts of infants and young children. Their primary role in disease is as the agents of urinary tract infection, infective endocarditis (like the viridans group streptococci), and wound infections, especially those contaminated with intestinal contents. In the laboratory, their colonies resemble somewhat those of group B streptococci. They must be differentiated from this organism because their presence at certain body sites has a different meaning. For example, enterococci isolated from a genital tract specimen of a pregnant woman near term may simply represent contamination from the intestinal tract, whereas isolation of group B streptococci from the same specimen represents a potential hazard for the fetus. *Enterococcus faecalis* is the most common species isolated from persons with enterococcal infections, but another species, *Enterococcus faecium*, is being isolated more frequently from hospitalized patients with serious infections. This organism is highly resistant to almost all antimicrobial agents including vancomycin, which has been the only drug available for treating some strains of this species. Enterococcal strains resistant to this agent are referred to as vancomycin-resistant enterococci, or VRE. Patients colonized or infected with VRE are treated with special precautions in the hospital to prevent transmission of the organism to others. Treatment of infections caused by VRE is a significant clinical challenge. Although pharmaceutical companies are working to develop new, effective drugs, microbial resistance to them evolves rapidly. Some strains of other enterococcal species, including *E. faecalis*, are also resistant to vancomycin but not yet to the same extent as *E. faecium* strains are.

Enterococci previously were known as group D streptococci because they possess a characteristic antigen on their cell wall that reacts in serological tests with group D antibody. Enterococcal colonies typically appear alpha-hemolytic, although they may be beta- or nonhemolytic, depending on the isolation medium used and incubation conditions. Unlike the streptococci, enterococci can grow in a high-concentration salt broth (containing 6.5% sodium chloride), are resistant to bile, and hydrolyze a complex carbohydrate, esculin. The last two characteristics are used in a selective and differential medium for enterococci, called bile-esculin agar. The bile inhibits streptococcal but not enterococcal growth. When enterococci hydrolyze the esculin, a black pigment forms in the medium. The pigment results from the reaction of the esculin breakdown products with an iron salt that is also included in the medium (see **colorplate 32**). This test often becomes positive within 4 hours so that a rapid identification can be made. An even more rapid test that is performed with colonies of enterococci growing on a culture plate is the PYR test. This test detects an enzyme, pyrrolidonylarylamidase, which is produced by enterococci but not most other gram-positive cocci (an important exception is the group A beta-hemolytic *Streptococcus*, which also produces pyrrolidonylarylamidase and is positive in the PYR test). The substrate for this enzyme is impregnated on disks and its hydrolysis is detected by a simple disk method (see **colorplate 33**).

Purpose	To identify enterococci in culture
Materials	6.5% sodium chloride broths
	Plate of bile-esculin agar
	PYR disks and developer reagent
	Blood agar plate cultures of *Enterococcus faecalis* and a group B *Streptococcus*
	Wire inoculating loop
	Forceps
	Glass microscope slide or empty Petri dish
	Distilled water
	Marking pen or pencil

Procedures

1. Examine the blood agar plate culture of the *Enterococcus*. Do the colonies resemble those of the group B *Streptococcus?* Make a Gram stain of the organisms.
2. Inoculate two labeled sodium chloride broths *lightly,* one each with a portion of a colony from each plate. After you inoculate them, the broths should not be turbid; otherwise, you will not be able to determine whether the organism grew during incubation.

3. Incubate the broths for 24 hours at 35°C.
4. Mark the bottom of the bile-esculin plate to divide it in half. Label one half "enterococcus" and the other "group B strep."
5. Streak the *Enterococcus* across one-half of the bile-esculin agar plate and the group B *Streptococcus* across the other half. Incubate the plate at 35°C and examine it just before you leave the laboratory (don't forget to reincubate) and again after 24 hours.
6. With forceps, remove a filter paper disk impregnated with PYR substrate (L-pyrrolidonyl-beta-naphthylamide) from the vial of disks. Place the disk on the surface of a glass microscope slide or in an empty Petri dish. Moisten the disk with a small drop of tap or distilled water, taking care not to flood the disk.
7. With your sterilized inoculating loop, pick up several colonies of enterococci and rub them onto the surface of the disk. Be careful not to dig up any blood agar with your inoculum. Resterilize your inoculating loop.
8. After 2 minutes, add a drop of the developer reagent to the surface of the disk. A red color develops within 1 minute if the test is positive.
9. Repeat steps 6 through 8 with the culture of group B *Streptococcus*.
10. After 24 hours, examine the salt broths for the presence or absence of growth (turbidity). Compare the inoculated, incubated broths with an uninoculated broth tube.
11. Examine the bile-esculin plate and note the color of the medium in each half.

Results

1. For each organism, record results (+ or −) of the esculin hydrolysis reaction (black pigment formation) at the end of the lab session (less than or equal to [≤] 4 hours) and at 24 hours.

   Enterococcus faecalis: ≤4 hours _____

        24 hours _____

   group B *Streptococcus:*   ≤4 hours _____

        24 hours _____

2. Record your observations in the following table.

| Name of Organism | Colonial Morphology | Gram-Stain Reaction and Microscopic Morphology | Bile-Esculin | | PYR | Salt Broth |
			Growth (+ or −)	Black Pigment (+ or −)	Red Color (+ or −)	Growth (+ or −)
Enterococcus faecalis						
Group B *Streptococcus*						

EXPERIMENT **20.5**   **Streptococci in the Normal Flora**

Purpose	To study the normal flora of the throat
Materials	Sheep blood agar plate
	Sterile swab
	Sterile tongue depressor
	Marking pen or pencil
	Optional:
	Simulated swab from suspected strep throat patient
	Kit for detection and confirmation of group A streptococcal antigen from a throat swab

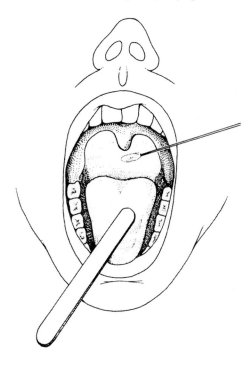

Procedures

1. Figure 20.3 diagrams the correct method for collecting a throat culture. Note that the tongue is held down out of the way and the throat swab is lightly touching the posterior wall of the pharynx. The tonsillar areas should also be sampled.
2. The instructor will demonstrate the method. Observe carefully.
3. Now take a throat culture from your laboratory partner. First label a blood agar plate with the person's name or initials. The "patient" should be positioned in good light so that you can see the back of the throat and the position of the swab as you insert it. *Gently* and *quickly* swab the posterior membranes of the throat, being careful not to touch the swab to any other tissues as you insert or remove it.
4. Inoculate the labeled blood agar plate by rolling the swab over a small area near one edge. Streak the plate with the inoculating loop in a manner to obtain isolated colonies (review fig. 7.1).
5. Discard the swab and tongue depressor in a container of disinfectant.
6. With your sterilized loop, make a few stabs in the agar at the area where the swab was rolled. Do not stab through to the bottom of the agar layer. Incubate the plate at 35°C for 24 hours.
7. If the group A streptococcal antigen test kit is available, take a second swab from your "patient" (step 3). Test both your "patient" swab and the simulated swab from the "strep" throat patient, following the manufacturer's instructions carefully.

Results

1. If the streptococcal antigen detection test was performed, record the results.

    Your laboratory partner "patient" (+ or −) _____

    "Strep" throat patient (+ or −) _____

Complete the following if the test you used was an EIA test.

Color of the positive test: _____

Color of the negative test: _____

2. Examine the incubated culture plate carefully. How many colonies of different types can you distinguish? Describe each colony type in the table below.
3. Hold the plate against a good light. Do you see any hemolytic colonies? Indicate the type of hemolysis shown by each colony recorded in the table.
4. Make a Gram stain of one colony of each type and record the results in the table.
5. Enter your tentative identification of each colony in the table and the additional tests needed to complete the identification.
6. Did the culture of your laboratory partner "patient's" throat confirm the results of the swab antigen detection test?

Colony Morphology	Type of Hemolysis	Gram-Stain Reaction and Microscopic Morphology	Tentative Identification	Further Tests Needed

Questions

1. Differentiate the microscopic morphology of streptococci and pneumococci as seen by Gram stain.

2. What type of hemolysis is produced by *S. pneumoniae?*

3. What is the quellung reaction?

4. What role does a bacterial capsule play in infection?

5. Why is blood agar considered a differential medium?

6. What is the function of a candle jar?

7. Describe the principle of the latex agglutination test.

8. Is the normal flora of the upper respiratory tract harmful to the human host? Explain.

9. Is the normal flora beneficial to the host? Explain.

10. In collecting a throat culture, why is it important not to touch the swab to other surfaces in the mouth?

11. What specimens are of value in making a laboratory diagnosis of bacterial pneumonia? Why? Explain the difference between saliva and sputum.

12. Would a direct Gram stain of a sputum specimen be of any immediate value to the physician in choosing treatment for a patient with pneumonia? Explain.

13. Does antimicrobial therapy have any effect on the body's normal flora? Explain.

14. What is the significance of VRE?

15. Name a vaccine that can be used to prevent serious pneumococcal infection.

Diagnostic Microbiology in Action

Haemophilus, Corynebacteria, and *Bordetella*

Learning Objectives

After completing this exercise, students should be able to:

1. Define X and V factors.
2. Name three species of *Haemophilus* and indicate the types of infection with which each may be associated.
3. List the media used to isolate *Haemophilus* species and their necessary components.
4. Name the etiologic agent of diphtheria and describe the media used to isolate it from a clinical specimen.
5. Name the etiologic agent of whooping cough and describe the media used to isolate it from a clinical specimen.

EXPERIMENT 21.1   *Haemophilus*

The genus *Haemophilus* contains a number of species of fastidious, gram-negative bacilli. Most of these are found as normal flora of the upper respiratory tract. *Haemophilus* species can cause infections in a variety of sites in the upper respiratory tract and elsewhere in the body. Laboratory diagnosis is made by identifying these organisms in clinical specimens that represent the area of infection (throat swab, sinus drainage, sputum, conjunctival swab, spinal fluid, blood, or other). A direct smear of the specimen may be useful in providing rapid, presumptive information, particularly for spinal fluid or an exudate from the eye. Smears of material from the upper respiratory tract, with its mixed flora, may have little value. Latex antibody tests can also be performed directly with certain patient body fluids to detect *Haemophilus* antigen (see Exercise 16). Until an effective vaccine came into widespread use in the early 1990s, most serious *Haemophilus* disease was caused by *H. influenzae* serogroup b (*H. influenzae* strains are divided into serogroups a–f on the basis of their antigenic polysaccharide capsule). This organism is seldom isolated in the clinical laboratory today, but other *Haemophilus* species and *H. influenzae* serogroups other than serogroup b are occasionally encountered.

The fastidious *Haemophilus* organisms require specially enriched culture media and microaerophilic incubation conditions. Chocolate agar is commonly used for primary isolation of *Haemophilus* from clinical specimens. Unlike blood agar, which contains intact red blood cells, this medium contains hemoglobin derived from bovine red blood cells as well as other enrichment growth factors. Because the hemoglobin is dark brown, the agar in the plate has the appearance of chocolate.

Two special growth factors, called X and V, are required by some *Haemophilus* species. Some require one but not the other. The X factor is *hemin,* a heat-stable derivative of hemoglobin (supplied in chocolate agar). The V factor is a heat-labile coenzyme (nicotinamide adenine dinucleotide, or NAD), essential in the metabolism of some species that lack it. Yeast extracts contain V factor and are one of the most convenient supplements of chocolate agar or other media used for *Haemophilus*. Organisms other than yeasts elaborate V factor. Staphylococci, for example, when growing on an agar plate, secrete NAD into the surrounding medium. *Haemophilus* species that need V factor may grow in the zone immediately around the staphylococci but not elsewhere on the plate. This growth of the dependent organism is described as "satellitism" (see **colorplate 34**). X and V factors can also be incorporated directly into agar media that do not contain these factors, or alternatively, they can be impregnated in filter-paper disks that are pressed on the surface of X and V factor–deficient media. In the latter case, the growth factors diffuse into the agar in a manner similar to diffusion from disks impregnated with antimicrobial agents (see Experiment 12.1).

Purpose	To identify *Haemophilus* species in culture
Materials	Sheep blood agar plates
	Chocolate agar plate
	Nutrient agar plate
	X and V disks
	Forceps
	Haemophilus ID Quad Plates
	Chocolate agar plate cultures of *Haemophilus influenzae* and *Haemophilus parainfluenzae*
	Demonstration blood agar and nutrient agar plates showing satellitism
	Wire or disposable inoculating loop
	Marking pen or pencil
	Candle jar or a CO_2 incubator

Procedures

1. Make a Gram stain of each species of *Haemophilus*.
2. Divide a sheep blood plate and a chocolate agar plate in half with your marking pen or pencil. Label one half *H. influenzae* and the other half *H. parainfluenzae*. Inoculate *H. influenzae* on the appropriate side of each plate and streak for isolation within this half. Repeat with *H. parainfluenzae* on the other half of each plate. Incubate these plates in a candle jar or CO_2 incubator at 35°C for 24 hours.
3. Repeat step 2 using the nutrient agar plate, but inoculate each strain heavily and streak for confluent growth within its half of the plate. Now, using heated, cooled forceps, place an X and a V disk on the agar surface streaked with *H. influenzae* and repeat on the *H. parainfluenzae* side. The two disks on each side should be placed not more than 1 inch apart, and centered in the area streaked (see diagram under Results, step 2). Incubate this plate in a candle jar or a CO_2 incubator at 35°C for 24 hours.
4. Label one Haemophilus ID Quad Plate with the name *H. influenzae*, and *lightly* streak all four quadrants of the plate with the *H. influenzae* culture. To prevent carryover of growth factors, sterilize the loop between inoculation of each quadrant or use a separate disposable loop for each inoculation. Repeat with a second plate using the *H. parainfluenzae* culture. Incubate these plates in a candle jar or CO_2 incubator at 35°C for 24 hours.
5. Examine the demonstration plates. *H. influenzae* has been streaked heavily on one half of each plate, *H. parainfluenzae* on the other half. An inoculum of a *Staphylococcus* culture was made in one area in the center of each streaked portion. Describe your observations and indicate your interpretation of the appearance of the blood and nutrient agar plates under Results, step 4.

Results

1. Complete the following table, describing the Gram-stain appearance of the two *Haemophilus* species and indicating any morphological distinctions you observed between them. Describe the colonial morphology of each *Haemophilus* species.

Organism	Gram-Stain Reaction and Microscopic Morphology	Colonial Morphology on	
		Chocolate Agar	Blood Agar
Haemophilus influenzae			
Haemophilus parainfluenzae			

2. Diagram the appearance of the growth of each *Haemophilus* species on the nutrient agar plate with X and V disks and interpret the growth requirements of each.

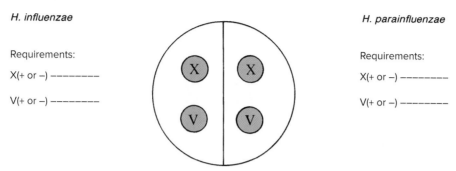

H. influenzae

Requirements:

X(+ or −) ――――――

V(+ or −) ――――――

H. parainfluenzae

Requirements:

X(+ or −) ――――――

V(+ or −) ――――――

3. Diagram the appearance of the growth of each *Haemophilus* species on the Quad Plates and interpret the nutritional requirements of each in the table below (see **colorplate 35**).

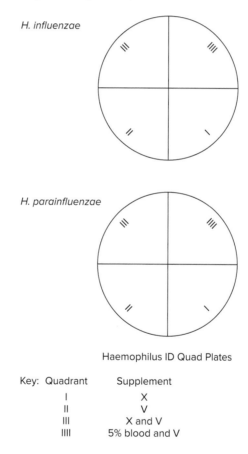

H. influenzae

H. parainfluenzae

Haemophilus ID Quad Plates

Key: Quadrant   Supplement
      I   X
      II   V
      III   X and V
      IIII   5% blood and V

Organism	Nutritional Requirement			
	X only (+ or −)	V only (+ or −)	X and V (+ or −)	5% Blood and V (+ or −)
Haemophilus influenzae				
Haemophilus parainfluenzae				

Haemophilus, Corynebacteria, and Bordetella

4. Diagram the appearance of the demonstration plates of satellitism and interpret.

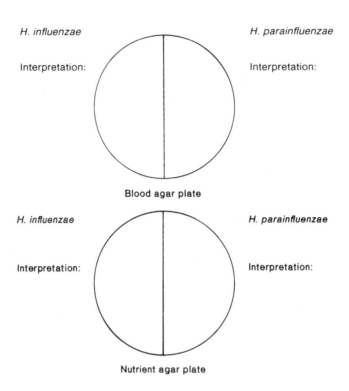

H. influenzae

Interpretation:

H. parainfluenzae

Interpretation:

Blood agar plate

H. influenzae

Interpretation:

H. parainfluenzae

Interpretation:

Nutrient agar plate

EXPERIMENT **21.2** **Corynebacteria**

The genus *Corynebacterium* is comprised of many species, but *Corynebacterium diphtheriae* has the most important pathogenic properties. *C. diphtheriae* is the agent of diphtheria, a serious throat infection and a systemic, toxic disease. If they have an opportunity to colonize in the throat, virulent strains of this organism not only damage the pharyngeal tissue, causing formation of a *pseudomembrane,* but they produce a powerful *exotoxin* that disseminates through the body from the site of its production in the upper respiratory tract. When this toxin reaches the heart, myocarditis and heart failure may result. Severe nervous system damage may cause paralysis of the limbs and diaphragm, and the kidney and other organs may also be affected. The systemic effect of the toxin is the primary cause of death in those patients with diphtheria who are not promptly recognized and treated. In rare cases, the skin rather than the throat is affected, but all toxic disease manifestations are the same. The disease is controlled by maintaining active immunization with diphtheria *toxoid* (purified toxin treated so that it is no longer toxic but remains immunogenic).

Early clinical and laboratory recognition of diphtheria infection developing in the throat is critical because prompt treatment with antitoxin (antibody that neutralizes the toxin) and an appropriate antimicrobial agent are required for patient recovery. In the laboratory, the microbiologist must distinguish *C. diphtheriae* from other corynebacteria that are harmless members of the normal flora but usually present in throat specimens. Identification must be made as rapidly as possible, for the laboratory report is essential for clinical decisions. In spite of widespread immunization in the United States, occasional sporadic outbreaks of both pharyngeal and skin diphtheria occur. Because respiratory diphtheria is so rare in the United States (0 to 5 cases per year), the limited epidemiologic, clinical, and laboratory expertise is a major challenge of the disease. Often, the clinician may not even consider this diagnosis. However, diphtheria remains endemic in developing countries. The countries of the former Soviet Union have reported more than 150,000 cases and 5,200 deaths in an epidemic that began in 1990. From 1980 to 2010, 55 cases of diphtheria in the United States were reported to the Centers for Disease Control and Prevention, including 4 fatal cases in unvaccinated children, but in the decade before 2016, fewer than 5 cases were reported.

Corynebacteria are gram-positive, nonmotile, nonsporing bacilli that, like staphylococci, are widely distributed on our bodies and in the environment. Nonpathogenic species are often called *diphtheroids* because their microscopic morphology resembles that of *C. diphtheriae*. These rods often contain granules that stain irregularly (they are said to be *metachromatic*) and give the organisms a beaded or clubbed appearance. Pairs or small groups characteristically fall into patterns that look like Chinese characters, or like Vs and Ys. Usually, *C. diphtheriae* is longer, thinner, and more beaded in appearance than diphtheroids, which are generally short and thick by comparison. This differentiation can be very difficult to make in examining a stained throat smear and cannot be relied on for accurate diagnosis. In patients with decreased immune function (referred to as *immunocompromised* patients), corynebacteria other than *C. diphtheriae* may cause disease by invading the weakened host to produce bacteremia and pneumonia.

In culture, corynebacteria are not highly fastidious. They grow well aerobically on nutrient media. When diphtheria is suspected, the primary isolation media used for throat swabs include those that are selective and differential for *C. diphtheriae* and also blood agar. Two media used for this purpose are cystine-tellurite blood agar and modified Tinsdale's agar, which should be incubated for a minimum of 48 hours. Colonies of *C. diphtheriae* on cystine-tellurite blood agar appear black or gray, whereas those on modified Tinsdale's agar are black with dark brown halos. The blackening of the medium around the colony is due to the organism's ability to reduce tellurite to telluride, which in the presence of cysteine in the medium, produces a black precipitate. Any suspected cases of diphtheria must be reported to local and state health departments, which would be involved in diagnosis and disease management.

To definitively identify a *C. diphtheriae* isolate as a true pathogen, demonstration of toxin production is required. In the past, a common test used for toxin detection was the Elek test. In this test, strips with antitoxin (antibody) against the *C. diphtheriae* toxin were placed on agar plates that were inoculated with the *C. diphtheria* isolate. If the antitoxin diffusing from the strip encountered the antigenic toxin produced by a toxigenic *C. diphtheriae* strain growing on the medium, a white line of antigen-antibody precipitate appeared in the agar. Currently, diphtheria antitoxin is available in the United States only through the Centers for Disease Control and Prevention. The polymerase chain reaction (PCR) and enzyme immunoassays are now used to detect *C. diphtheriae* toxin production.

Purpose	To identify corynebacteria in smears and cultures
Materials	Blood agar plate
	Blood tellurite plate
	Tubed phenol red glucose broth
	Tubed phenol red maltose broth
	Tubed phenol red sucrose broth
	Methylene blue
	Prepared Gram- and methylene-blue-stained smears of *C. diphtheriae*
	Loeffler's slant cultures of *Corynebacterium xerosis* and *Corynebacterium pseudodiphtheriticum*
	Nutrient agar slant culture of *Escherichia coli*
	Wire inoculating loop
	Marking pen or pencil

Procedures

1. Prepare a Gram stain and a methylene blue stain (see Exercises 3 and 4) from either one of the *Corynebacterium* cultures. Read and compare these with the prepared Gram- and methylene-blue-stained smears of *C. diphtheriae,* recording your observations under Results.
2. Inoculate a blood agar plate with either one of the *Corynebacterium* cultures. Streak for isolation.
3. Divide the blood tellurite plate into two parts with your marker and label each half appropriately. Inoculate one side of the plate with a *Corynebacterium* species, the other side with *E. coli.*
4. Inoculate the *C. xerosis* culture into each of the three carbohydrate broths. Repeat with the culture of *C. pseudodiphtheriticum.*
5. Incubate all plate and tube cultures at 35°C for 24 hours.
6. Examine your cultures and record your observations.

Results

1. Illustrate the microscopic morphology of corynebacteria:

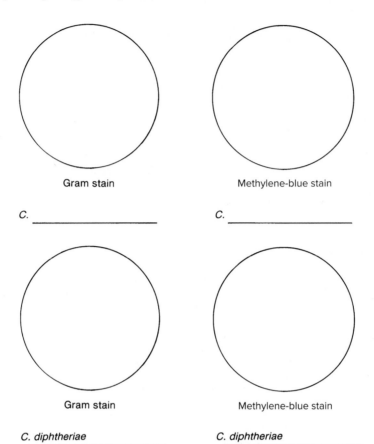

Gram stain

Methylene-blue stain

C. _____

C. _____

Gram stain

Methylene-blue stain

C. diphtheriae

C. diphtheriae

2. Describe the appearance of a *Corynebacterium* species on blood agar.

3. Describe the appearance of *E. coli* and of a *Corynebacterium* species on blood tellurite agar.

4. Complete the following table.

Name of Organism	Glucose	Maltose	Sucrose
Corynebacterium xerosis			
Corynebacterium pseudodiphtheriticum			
Corynebacterium diphtheriae*			

*Results from your reading.

EXPERIMENT **21.3** *Bordetella*

Bordetella pertussis is the etiologic agent of whooping cough. This severe upper respiratory tract infection was common (more than 200,000 cases per year) and affected children primarily until the 1940s, when whooping cough vaccine became widely administered. It is clear, however, that childhood immunization does not provide life-long immunity, and pertussis-like infection is now seen more frequently in adolescents and adults than in children. As of the middle of July 2012, nearly 18,000 cases of whooping cough and 9 deaths were reported in that year—a 50-year high in the United States. A less-effective, newly modified acellular version of the pertussis vaccine may be partly responsible for the increase in cases. As a result, the Centers for Disease Control and Prevention has recommended that adolescents and adults ages 11 to 64 receive a booster dose of pertussis vaccine at the time they receive their diphtheria and tetanus boosters.

    Bordetella pertussis is a fastidious organism that grows best on special media. The two most common are Bordet-Gengou (BG) agar, which is enriched with glycerin, potato, and 30% defibrinated sheep blood, and Regan-Lowe (RL) agar, which consists of charcoal agar, defibrinated horse blood, and an antimicrobial agent, cephalexin, to inhibit growth of normal respiratory flora. The charcoal is present to adsorb toxic substances that might be present in the agar. Visible colonies are produced only after 3 to 5 days incubation in a microaerophilic atmosphere. On BG medium, the colonies are raised, rounded, and glistening (resembling mercury droplets or a bisected pearl), and usually have a hazy zone of hemolysis. On RL medium, the colonies are round, domed, shiny, and may run together slightly.

    B. pertussis is a gram-negative bacillus resembling *Haemophilus* species, with which it was once classified. When whooping cough is suspected, the best specimen for laboratory diagnosis is a nasopharyngeal swab, but throat swabs may be used in addition. Although isolation of *B. pertussis* in culture is the diagnostic test in the laboratory, it is a relatively insensitive method because of the organism's fastidious growth requirements. The Centers for Disease Control and Prevention now recommend that PCR assays be used for diagnosis when available, because they are more sensitive than culture.

Purpose	To observe *Bordetella pertussis* in demonstration and to examine a throat culture on Bordet-Gengou (BG) and Regan-Lowe (RL) media
Materials	Prepared Gram stains of *B. pertussis* Projection slides, if available Bordet-Gengou and Regan-Lowe agar plates Sterile swab Marking pen or pencil Candle jar or a CO_2 incubator

Procedures

1. Examine the prepared Gram stains and record your observations.
2. Observe colonial morphology as demonstrated.
3. Collect a throat specimen as in Experiment 20.5 and inoculate the Bordet-Gengou and Regan-Lowe plates. Label the plate with the patient's name and incubate the plates at 35°C in a candle jar or CO_2 incubator for 24 hours. If no growth is observed, reincubate the plates and observe periodically for up to 7 days.

Results

1. Describe the Gram-stain reaction and microscopic morphology of *B. pertussis*.

2. Describe your observations of the demonstration material.

3. Describe the appearance of colonies on your Bordet-Gengou and Regan-Lowe throat culture plates.

What is the total number of colonies? BG _____ RL _____

How many colony types can be distinguished? BG _____ RL _____

How does the flora compare with that of your throat culture in Experiment 20.5?

CASE STUDY

A School-Associated Pertussis Outbreak

A 13-year-old eighth grader attended school while experiencing a prolonged cough illness. *Bordetella pertussis* was isolated in culture from the child. Another student in the same classroom had a clinical illness consistent with pertussis 2 weeks earlier, and subsequently, five additional people (two students in the same classroom, two eighth-grade teachers, and one parent of an ill student) developed the coughing illness. In patients with one or more days of illness, the diagnosis was made by culturing the organism from a nasopharyngeal swab. In patients with more than 14 days of illness, the diagnosis was made by either a positive PCR test for *B. pertussis* DNA from a nasopharyngeal specimen or by finding a link to a person with a laboratory-confirmed case. Antimicrobial therapy was given to students and staff members with coughing, and they were excluded from school through the fifth day of treatment.

      Other cases began appearing throughout six communities in the county. As a result, the pertussis vaccination schedule for infants was accelerated. A total of 485 pertussis cases were reported throughout the outbreak, which lasted 6 months. Genetic testing revealed four DNA profiles of *B. pertussis,* most of which matched the profile of patients attending the middle school. No patients were hospitalized or died during this outbreak. Early recognition, treatment, and chemoprophylaxis were important in preventing pertussis transmission to others.

1. Would you guess that diagnosis would have been improved if throat swabs rather than nasopharyngeal swabs were cultured or examined for *B. pertussis* DNA?
2. What are two laboratory methods that can be used to diagnose pertussis?
3. What is the recommended age range for receiving the pertussis vaccine?
4. Do determining genetic profiles help in studying disease transmission? How?

Source: Adapted from S. Everett et al. 2004. School-Associated Pertussis Outbreak—Yavapai County, Arizona, September 2002–February 2003. *MMWR Weekly* 53:216–219.

Diagnostic Microbiology in Action

Questions

1. What is the satellite phenomenon?

2. What is the incidence of *Haemophilus influenzae* as an agent of meningitis in infants and children under 3 years of age? In adults?

3. Why is a direct smear of spinal fluid essential when bacterial meningitis is suspected?

4. How can a diphtheroid be distinguished from the agent of diphtheria?

5. What is a diphtheria toxin test and how is it performed?

6. Can diphtheria be transmitted directly via the respiratory route? If so, how?

7. How is diphtheria prevented?

8. Why is early laboratory diagnosis of diphtheria important?

9. What is the preferred specimen for diagnosing whooping cough?

10. How can transmission of respiratory infections be prevented?

Diagnostic Microbiology in Action

11. Complete the following table.

Bacteria Associated with the Respiratory Tract and with Disease

Etiologic Agent	Disease	Specimens for Lab Diagnosis	Microscopic Morphology and Gram-Stain Reaction	Hemolysis (Type)	Key Tests for Lab Identification	Normal Habitat
S. pyogenes						
Alpha-hemolytic streptococci						
S. pneumoniae						
E. faecalis						
S. epidermidis						
S. aureus						
C. diphtheriae						
Diphtheroids						
H. influenzae						
H. haemolyticus						
B. pertussis						

Clinical Specimens from the Respiratory Tract

Learning Objectives

After completing this exercise, students should be able to:
1. Explain the importance of group A streptococci as a cause of sore throat.
2. List the immune complex diseases that can follow group A *Streptococcus* infection.
3. Discuss the value of the Gram-stained smear in evaluating the quality of a sputum specimen.
4. Understand how a Gram-stained smear can be used to make a presumptive diagnosis of pneumonia.
5. Explain the importance of performing an antimicrobial susceptibility test on certain bacteria recovered from clinical samples.

Now that you have had some experience with the normal flora and the most common bacterial pathogens of the respiratory tract, you will have an opportunity to apply what you have learned to the laboratory diagnosis of respiratory infections. In Experiments 22.1 and 22.2, you will prepare cultures of a throat swab and a sputum specimen, each simulating material that might be obtained from a sick patient. These cultures should be examined with particular attention to the "physician's" stated tentative diagnosis. Significant organisms that may be isolated must be identified and reported. If organisms that you consider to be part of the normal flora are isolated, report as "normal flora."

In Experiment 22.3, you will set up an antimicrobial susceptibility test on an organism isolated from one of the clinical specimens previously cultured and prepare a report of the results for the "physician."

EXPERIMENT **22.1**   **Laboratory Diagnosis of a Sore Throat**

The group A *Streptococcus* (*Streptococcus pyogenes*) is the leading cause of bacterial pharyngitis (sore throat). If the infection is not treated properly with antibiotics, the patient may develop serious associated conditions following infection. These conditions are thought to be immunologic diseases, known as immune complex diseases. In patients with group A streptococcal pharyngitis, antibodies produced against the streptococci during infection react with heart valve or kidney tissue, resulting in rheumatic fever or glomerulonephritis, respectively. Certain viruses also cause sore throat, but they are not grown by using bacterial culture techniques and do not cause immune complex diseases. In order to administer appropriate antibiotic therapy promptly, the physician must know whether the group A *Streptococcus* is responsible for the patient's sore throat. Prompt treatment avoids the production of significant amounts of antibodies against group A streptococci and prevents the development of rheumatic fever and glomerulonephritis.

Purpose	To identify bacterial species in a simulated clinical throat culture as quickly as possible
Materials	Swab in a tube of broth, accompanied by a laboratory request for culture
	Patient's name: Mary Peters
	Age: 6 years
	Physician: Dr. M. Selby
	Tentative clinical diagnosis: Strep throat
	Blood agar plate (BAP)
	Forceps
	PYR disks and developer reagent
	Tubes containing 0.4 ml streptococcal extraction enzyme or prepared extract
	Capillary pipettes
	Latex test kit for serological typing
	Wire inoculating loop
	Marking pen or pencil
	Test tube rack
	Water bath

Procedures

1. Using the swab in the "specimen" tube, inoculate a small area of the blood agar plate labeled "specimen" on the bottom. Discard the swab in disinfectant solution or a biohazard container. With a sterilized inoculating loop, streak the remainder of the plate to obtain isolated colonies. After you have completed the streaking step, make a few shallow cuts with your loop in the area of the original inoculum.
2. Incubate the plate at 35°C for 24 hours.
3. After the plate has incubated, examine it carefully for the presence of hemolysis and record the type of hemolysis you see on the laboratory work card (page 183). Record the colonial morphology and make Gram stains of different colony types.
4. On the basis of your findings, record on the Microbiology Laboratory Report (page 183) the preliminary result that you will give to the "physician" when he or she calls for a report.
5. With your sterilized inoculating loop, perform a PYR test by rubbing several beta-hemolytic colonies with similar morphologies onto a moistened disk impregnated with PYR reagent. After 2 minutes, add a drop of the developer reagent to the surface of the disk. The appearance of a bright red color within 5 minutes signifies a positive test for group A streptococci. In a negative test, there is either no color change or an orange color develops (see **colorplate 33**). Record your results on the laboratory work card.
6. If the instructor has not prepared an extract of beta-hemolytic colonies grown from the simulated throat specimen, follow steps 7 and 8.
7. With your sterile loop, make a light suspension of "suspicious" beta-hemolytic colonies in 0.4 ml of extraction enzyme. Five or six colonies should be sufficient.
8. Place the suspension in a 35°C water bath or in a beaker of water warmed to 35°C in an incubator. After 5 minutes, shake the tube and continue incubating for no less than 10 minutes and up to 1 hour.
9. Following the manufacturer's directions, mix one drop of group A latex reagent and one drop of group B latex reagent each with a drop of your extract on a glass slide or special reaction card provided. Rock the slide back and forth for at least 1 minute, looking for the formation of agglutinated latex particles and a clearing of the background (see fig. 20.1).
10. If agglutination is present, record the group (A or B) on your work card along with the final organism identification(s).
11. Complete the Microbiology Laboratory Report for the "physician."

Results

1. Laboratory work card (record of your work to be kept on file for at least 2 years).

Culture No.:		Patient's Name:		Physician:		
Specimen Type:		Date Received:		Date Reported:		
Colonial Morphology	Gram-Stain Reaction and Morphology	Type of Hemolysis	PYR Test (+ or –)	Group (by Latex)	Name of Organism	
Final Report:			Signature:			

2. Final laboratory report to "physician."

MICROBIOLOGY LABORATORY REPORT

Patient's Name: _____

Sex: _____ Age: _____ Date: _____

Tentative Diagnosis: _____

Laboratory Findings

Preliminary Culture Result: _____

Final Culture Result: _____

SIGNATURE: _____ Date Received: _____

LABORATORY NAME: _____

PHYSICIAN'S NAME: _____

Date Reported: _____

EXPERIMENT 22.2   Laboratory Diagnosis of Bacterial Pneumonia

The laboratory diagnosis of bacterial pneumonia can be problematic because some bacteria that are the most common causes of pneumonia (*Streptococcus pneumoniae, Haemophilus influenzae, Staphylococcus aureus,* and *Neisseria meningitidis*) are normal flora of the upper respiratory tract. The specimen of choice for establishing the laboratory diagnosis of pneumonia is sputum, which represents secretions from the lower respiratory tract. However, when sputum is coughed up or expectorated, it can become contaminated by saliva and its normal flora from the upper airway. In this instance, it is difficult to determine the clinical significance of the isolate growing in culture. The question is whether the isolate represents the cause of pneumonia or is merely contamination from upper respiratory tract secretions. To minimize this problem, many clinical microbiology laboratories routinely perform a Gram-stained smear of expectorated sputum samples. The Gram-stained smear can be performed quickly and inexpensively. It provides invaluable information about the quality of the sample and the predominant morphology of the bacterial types (also known as morphotypes) present. For example, the presence of many epithelial cells from the mucous membranes of the upper respiratory tract suggests a poor-quality sample that was heavily contaminated with saliva during collection. In contrast, the presence of many inflammatory cells (neutrophils) indicates that the sample is of high quality and that an infectious process is present in the lung. Finally, the Gram-stained sputum smear can provide rapid information about the predominant bacterial morphotype present, for example, gram-positive diplococci suggestive of *Streptococcus pneumoniae*. This information can be used to establish a presumptive diagnosis of pneumococcal pneumonia and guide the choice of empiric antibiotic therapy until the confirmatory microbiology culture report is received, perhaps after several days. Similarly, the presence of gram-positive cocci in clusters along with neutrophils can provide an early presumptive diagnosis of *Staphylococcus aureus* pneumonia (see **colorplate 1**, right).

Purpose	To identify bacterial species in a simulated sputum as quickly as possible
Materials	Simulated sputum in a screw-cap container, accompanied by a laboratory request for culture
	Patient's name: Richard Wilson
	Age: 72 years
	Physician's name: Dr. F. Smythe
	Tentative diagnosis: lobar pneumonia
	Wire inoculating loop
	Blood agar plate (BAP)
	Mannitol salt agar plate (MSA)
	Dropping bottle containing 3% hydrogen peroxide
	Tubed plasma (0.5-ml aliquots)
	Latex agglutination kit for *Staphylococcus aureus*
	Sterile 1.0-ml pipettes
	Pipette bulb or other aspiration device
	Marking pen or pencil

Procedures

1. Make a Gram stain of the simulated sputum specimen. Record the results and place the information on your work card (page 185).
2. Label a blood agar and a mannitol salt agar plate with the patient's name, and with your sterilized inoculating loop, inoculate each plate from the specimen. Streak each for isolation of colonies. Incubate both plates at 35°C for 24 hours.
3. When the "physician" calls, refer to your work card and give him or her specific information about your microscopic interpretation of the Gram-stained smear.
4. After the plates have incubated, examine each carefully. Record colonial morphology on the work card, and make Gram stains of different colony types on each medium.
5. Perform the catalase test on different colony types on each medium. Be careful not to scrape the surface of the blood agar plate or a false-positive reaction will occur, because red blood cells contain the enzyme catalase.
6. Perform the rapid latex agglutination test with any colony from either culture plate that resembles a *Staphylococcus* colony. Place a small drop of saline onto the test card, and with your sterile inoculating loop, prepare a suspension by rubbing the organism into the drop with a circular motion. Add a drop of the latex reagent and gently rotate and rock the card for 60 seconds. The appearance of an agglutination reaction signifies a positive reaction for *Staphylococcus aureus* (see fig. 19.1).
7. Alternatively, if the latex test is not available, perform the coagulase test with any colony on either plate that appears to be a *Staphylococcus*. With a sterilized inoculating loop, pick up a colony and emulsify it directly in 0.5 ml of plasma. Incubate the plasma tube and read at 30 minutes. If necessary, incubate the tube overnight and read the result the next day. Record the result on your work card.
8. Prepare a final report for the "physician."

Results

1. Laboratory work card (record of your work to be kept on file at least 2 years).

Culture No.:		Patient's Name:			Physician:		
Specimen Type:		Date Received:			Date Reported:		
Colonial Morphology on Blood Agar Plate	Gram-Stain Reaction and Morphology on Blood Agar Plate	Type of Hemolysis on Blood Agar Plate	Mannitol Reaction on Mannitol Salt Agar Plate (+ or −)	Catalse (+ or −)	Latex Agglutination Test (+ or −)	Coagulase Test (+ or −)	Name of Organism

2. Final laboratory report to "physician."

MICROBIOLOGY LABORATORY REPORT

Patient's Name: _____

Sex: _____ Age: _____ Date: _____

Tentative Diagnosis: _____

Laboratory Findings

Direct Smear Report: _____

Final Culture Result: _____

SIGNATURE: _____ Date Received: _____

LABORATORY NAME: _____

PHYSICIAN'S NAME: _____

Date Reported: _____

EXPERIMENT 22.3   Antimicrobial Susceptibility Test of an Isolate from a Clinical Specimen

Once a bacterium that may be the cause of the patient's infection has been isolated from a clinical specimen and identified, it may be necessary to determine its antimicrobial susceptibility. In this way, the physician knows which antibiotics may be effective in treating the patient's infection. This information is critical, because some bacteria may be resistant to the antibiotic being administered before the laboratory report is received, and the infection would not be eradicated. The antibiotic susceptibility report is one of the most important pieces of information provided by the clinical microbiology laboratory.

Purpose	To determine the antimicrobial susceptibility pattern of an organism isolated from a clinical specimen (in Experiment 22.2)
Materials	Nutrient agar plates (Mueller-Hinton, if available)
	Antimicrobial disks
	Sterile swabs
	Forceps
	Blood agar plate with pure culture of isolate
	Tube of nutrient broth (5.0 ml)
	McFarland No. 0.5 turbidity standard
	Marking pen and pencil

Procedures

1. Using a sterile swab, take some of the growth of a pure culture you isolated from the clinical specimen in Experiment 22.2, and emulsify it in 5.0 ml of nutrient broth until the turbidity is equivalent to the McFarland 0.5 standard. Discard the swab.
2. Take another sterile swab, dip it in the broth suspension, drain off excess fluid against the inner wall of the tube.
3. Inoculate a labeled agar plate as described in Experiment 12.1.
4. Follow procedures 4 through 7 of Experiment 12.1.
5. Incubate the agar plate at 35°C for 24 hours.
6. Examine plates and record results for each antimicrobial disk as S (susceptible), I (intermediate), or R (resistant).
7. Prepare a report for the "physician."

Results

Record results:

MICROBIOLOGY LABORATORY REPORT

Patient's Name: _____

Sex: _____ Age: _____ Date: _____

Tentative Diagnosis: _____

Antimicrobial Susceptibility Report

Name of Organism: _____

Source: _____

Antimicrobial Agent	S	I	R	Antimicrobial Agent	S	I	R

SIGNATURE: _____ Date Rec'd.: _____ Date Reported: _____

LABORATORY NAME: _____

PHYSICIAN'S NAME: _____

Diagnostic Microbiology in Action

Questions

1. Is a Gram stain of a throat swab useful for making a rapid, presumptive diagnosis of strep throat? Why?

2. Is a Gram stain of a sputum specimen useful in making a rapid, presumptive diagnosis of pneumonia?

3. How is a Gram-stained smear of sputum useful for evaluating the quality of the specimen?

4. What is the clinical significance of staphylococci isolated from throat specimens?

5. What is the clinical significance of staphylococci isolated from sputum specimens?

6. What is the clinical significance of beta-hemolytic streptococci isolated from throat specimens?

7. In a Gram stain of a sputum specimen, which type of body cell provides an indication that the specimen represents material from an active infection? Why?

8. Should an antimicrobial susceptibility test be performed on every bacterium isolated from a clinical specimen?

9. Which two immune complex diseases may develop following group A streptococcal infections that are not properly diagnosed and treated?

The *Enterobacteriaceae* (Enteric Bacilli) and Other Clinically Important Gram-Negative Bacilli

Learning Objectives

After completing this exercise, students should be able to:

1. Specify the types of bacteria found as normal flora in the human intestinal tract.
2. Explain the basis of biochemical and serologic tests for the characterization of enteric bacteria.
3. List the major components of TSI agar and explain the basis of the various test reactions.
4. Prepare a basic flow chart for isolating enteric bacteria from fecal samples.
5. Define and provide an example of an opportunistic infection and an opportunistic pathogen.

The human intestinal tract is inhabited soon after birth by a variety of microorganisms acquired, at first, from the mother. Later, organisms are carried in with food and water or introduced by hands and other objects placed in the mouth. Once ingested, many microorganisms cannot survive the acid conditions encountered in the stomach or the activity of digestive enzymes in the upper part of the intestinal tract. The small intestine and lower bowel, however, offer appropriate conditions for survival and multiplication of many microorganisms, primarily *anaerobic* species, that live there in extremely large numbers without harming their host.

When feces are cultured on bacteriologic media, it becomes apparent that most *facultatively anaerobic* bacterial species normally inhabiting the intestinal tract are gram-negative, nonsporing bacilli with some culture characteristics in common. This group of organisms is known as "enteric bacilli," or, in taxonomic terms, the family *Enterobacteriaceae*. However, some of the bacterial species that are classified within this group are important agents of intestinal disease. These usually are acquired through ingestion and are referred to as "enteric pathogens." The anaerobic organisms play little role in enteric disease and are not recovered in routine fecal cultures because they require special techniques for isolation (see Exercise 28).

One enteric organism that normally inhabits the intestinal tract, *Klebsiella pneumoniae,* is also sometimes associated with pneumonia. It is a gram-negative, nonmotile bacillus (see **colorplate 4**) that can cause infection when it finds an opportunity to invade the lungs or other soft tissue and the bloodstream. Like the pneumococcus, pathogenic strains of *K. pneumoniae* possess a slimy, protective capsule that is larger and more pronounced than most bacterial capsules (see **colorplate 12**).

In the experiments of this exercise we shall first study some of the cultural characteristics of those enteric bacilli that normally inhabit the bowel, and then apply this knowledge to understanding the methods used for isolating and identifying the important enteric pathogens.

The gram-negative enteric bacilli are not fastidious organisms. They grow rapidly and well under aerobic conditions on most nutrient media. The use of selective and differential culture media plays a large role in their isolation and identification. Their response to suppressive agents incorporated in culture media and their specific use of carbohydrate or protein components in the media provide the key to sorting and identifying them (review Exercises 13–15). A final identification by serological means can also be made as performed in Experiment 23.4.

EXPERIMENT **23.1**   **Identification of Pure Cultures of *Enterobacteriaceae* from the Normal Intestinal Flora**

Purpose	To learn how enteric bacilli are identified biochemically
Materials	Slants of triple-sugar iron agar (TSI)
	Methyl-red Voges-Proskauer (MR-VP) broths
	Slants of Simmons citrate agar
	Urea broths
	Slants of phenylalanine agar
	Lysine and ornithine decarboxylase broths
	Mineral oil in dropper bottle
	Sterile 1.0-ml pipettes
	Pipette bulb or other aspiration device
	Sterile empty test tubes and test tube rack
	Cotton-tipped swabs
	Kovac's reagent
	Methyl red indicator
	5% alphanaphthol
	40% potassium hydroxide
	10% ferric chloride
	Nutrient agar slant cultures of *Escherichia coli, Citrobacter koseri, Klebsiella pneumoniae,* pigmented and nonpigmented *Serratia marcescens, Enterobacter aerogenes, Proteus vulgaris,* and *Providencia stuartii*
	Marking pen or pencil
	Test tube rack

Procedures

1. Each student will be assigned two of the nutrient agar slant cultures. Label a set of culture media with the name of each organism. Inoculate each culture into the following media:
   TSI (using a straight wire inoculating needle; stab the butt of the tube and streak the slant; the closure should not be tight)
   MR-VP broth
   Simmons citrate agar slant
   Urea broth
   Phenylalanine agar slant
   Lysine decarboxylase broth
   Ornithine decarboxylase broth
2. Carefully overlay the surfaces of the lysine and ornithine broths with 1/2 inch of mineral oil.
3. Incubate subcultures at 35°C for 24 hours. Simmons citrate agar, MR-VP, and the decarboxylation assays may require incubation for 5 to 7 days before reactions are evident.
4. Before returning to class, read the following descriptions of the biochemical reactions to be observed and instructions for performing them.

Biochemical Reactions and Principles

A. TSI.   TSI contains glucose, lactose, and sucrose as well as a pH-sensitive color indicator. It also contains an iron ingredient for detecting hydrogen sulfide production, which blackens the medium if it occurs. TSI is similar to Kligler iron agar, which was described in Exercise 14 and is illustrated in **colorplate 19.**

Fermentation of the sugars by the test organism is interpreted by the color changes in the butt and the slant of the medium (see following table).

Butt	Slant	
Color*	Color*	Interpretation
Yellow	Yellow	Glucose and lactose, and/or sucrose fermented
Yellow	Orange-red or pink	Glucose only fermented
Orange-red	Orange-red	No fermentation
Bubbles		Gas production
Black		Hydrogen sulfate production

*Yellow signifies acid production, orange-red a neutral or negative reaction, and pink an alkaline reaction (breakdown of protein rather than carbohydrate).

B. IMViC Reactions.

The term **IMViC** is a mnemonic for four reactions: the letter **I** stands for the *indole test,* **M** for the *methyl red test,* **V** for the *Voges-Proskauer* reaction (with a small *i* added to make a pronounceable word), and **C** for *citrate.*

The *indole test* for tryptophan utilization was described in Experiment 14.3. Perform it in the same way here, using Kovac's reagent and a cotton-tipped swab.

Methyl red is an acid-sensitive dye that is yellow at a pH above 4.5 and red at a pH below 4.5. When the dye is added to a culture of organisms growing in glucose broth, its color indicates whether the glucose has been broken down completely to highly acidic end products with a pH below 4.5 (methyl red *positive*, red), or only partially to less acidic end products with a pH above 4.5 (methyl red *negative*, yellow).

The *Voges-Proskauer test* can be performed on the same glucose broth culture used for the methyl red test (MR-VP broth). One of the glucose fermentation end products produced by some organisms is acetylmethylcarbinol. The VP reagents (alphanaphthol and potassium hydroxide solutions) oxidize this compound to diacetyl, which in turn reacts with a substance in the broth to form a new compound having a pink to red color. VP-*positive* organisms are those reacting in the test to give this pink color change.

To perform the MR and VP tests, first withdraw 1.0 ml of the MR-VP broth culture, place this in an empty sterile tube, and set the tube aside for the VP test. Discard the pipette in disinfectant.

Do a methyl red test by adding 5 drops of methyl red indicator to 5.0 ml of MR-VP broth culture. Observe and record the color of the dye.

Perform a VP test by adding 0.6 ml of alphanaphthol and 0.2 ml of KOH solutions to 1.0 ml of MR-VP broth culture. Shake the tube well and allow it to stand for 10 to 20 minutes. Observe and record the color.

Citrate can serve some organisms as a sole source of carbon for their metabolic processes, but others require organic carbon sources. The citrate agar used in this test contains bromthymol blue, a dye indicator that turns from green to deep blue in color when bacterial growth occurs. If no growth occurs, the medium remains green in color and the test is negative.

C. H_2S Production.

This property is observed in TSI, as described in section A.

D. Urease Production.

The test for urease was described in Experiment 15.1. Read and record the results of your cultures tested in urea broth.

E. Phenylalanine Deaminase (PD).

The test for production of this enzyme was described in Experiment 15.5. Perform it in the same way, adding ferric chloride solution to your cultures on phenylalanine agar medium.

F. Lysine (LD) and Ornithine (OD) Decarboxylases.

Lysine and ornithine are amino acids that can be broken down by decarboxylase enzymes possessed by some bacteria. During this process of decarboxylation, the carboxyl (COOH) group on the amino acid molecule is removed, leaving alkaline end products that change the color of the pH indicator. In the broth test you use, a positive test is a deep purple color; a negative test is yellow. The reactions work best when air is excluded from the medium; therefore, the broths are layered with mineral oil after inoculation and before incubation.

Results

Record results for your cultures in the following table. Obtain results for other cultures by observing those assigned to fellow students.

Genus of Organism		TSI											
		Slant*	Butt*	I	M	Vi	C	H$_2$S	Urease†	PD	LD	OD
Escherichia													
Citrobacter													
Klebsiella													
Enterobacter													
Serratia‡	1												
	2												
Proteus													
Providencia													

*A = acid; K = neutral or alkaline; G = gas.
†If positive, specify time.
‡1 = pigmented strain; 2 = nonpigmented strain.

EXPERIMENT 23.2 Isolation Techniques for Enteric Pathogens

Bacterial diseases of the intestinal tract can be highly communicable and may spread in epidemic fashion. The agents of these diseases enter the body through the mouth in contaminated food or water, or as a result of direct contacts with infected persons. Among the *Enterobacteriaceae,* the organisms of pathogenic significance belong to the genera *Salmonella, Shigella,* and *Yersinia.* Also certain *Escherichia coli* strains can produce disease by several mechanisms, including invading tissue or producing toxins. Such strains are referred to as enteroinvasive or enterotoxigenic, respectively. Within the last 30 years, a new strain of *E. coli* has emerged as a significant human pathogen. This strain can cause bloody diarrhea and hemolytic uremic syndrome, which is especially severe and can be fatal in young children. Often referred to as enterohemorrhagic or Shiga toxin-producing *E. coli,* this strain has been responsible for sporadic outbreaks of disease associated with the ingestion of improperly cooked hamburger, unpasteurized cider, and fecally contaminated, uncooked vegetables and fruits.

In current nomenclature, the many former species of *Salmonella* are now referred to as *Salmonella* serotypes. They can be distinguished on the basis of their serological properties and certain biochemical activities. These organisms characteristically cause acute gastroenteritis when ingested, but some also can find their way into other body tissues and cause systemic disease. Among these, the most important is *Salmonella* serotype Typhi, the agent of typhoid fever, a serious systemic infection. The salmonellae are gram-negative bacilli that are usually motile. They usually do not ferment lactose but display a variety of other fermentative and enzymatic activities.

Shigella species are the agents of bacillary dysentery. These organisms are gram negative and nonmotile. They usually do not ferment lactose. In fermenting other carbohydrates, they produce acid but not gas (with one exception). They can also be identified to the level of species by serological methods.

Yersinia enterocolitica is the cause of acute enterocolitis, primarily in children. Its symptoms may mimic those produced by *Salmonella, Shigella,* or enteroinvasive *E. coli.* Occasionally, the symptoms are more suggestive of acute appendicitis. The organism grows better at room temperature (25°C) than at 35°C; therefore, it may not be isolated unless the physician notifies the laboratory that yersiniosis is suspected. In this case the isolation plates are incubated at both temperatures. *Yersinia* are gram-negative bacilli that are motile at 25°C but not at 35°C. They ferment sucrose, but not lactose. *Y. enterocolitica* is urease positive.

Disease-producing *E. coli* were once thought to be associated only with epidemic diarrhea in babies, but they are now known to be a common cause of "traveler's diarrhea" ("turista") and a variety of other gastrointestinal diseases. Some of

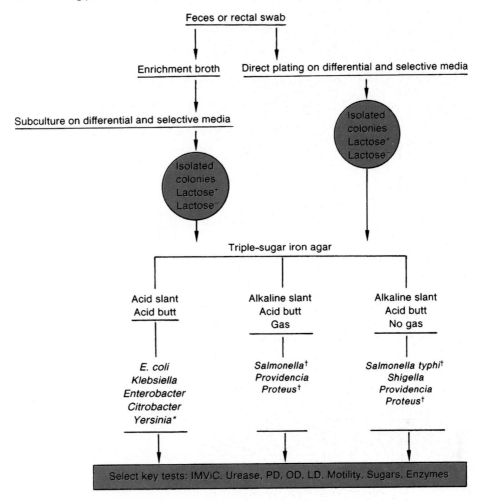

*Although *Yersinia* is lactose negative, it is sucrose positive.
†H_2S produced.

these strains may be distinguished from others by immunological typing of cell wall (O) and flagellar (H) antigens. In addition, an enzyme immunoassay is available to detect *E. coli* toxins directly in stool specimens.

The pathogenic *Enterobacteriaceae* are first isolated from clinical specimens by using highly selective media to suppress the normal flora in feces and to allow the pathogens to grow. Many of these media contain lactose, with a pH indicator, to differentiate the lactose-nonfermenting *Salmonella, Shigella,* and *Yersinia* (colorless on these agars) from any lactose-fermenting normal flora that may survive (pink-red or yellow colonies; see **colorplates 16** and **17**). Eosin methylene blue (EMB) or MacConkey agar is commonly used, together with two more highly selective media such as Hektoen enteric (HE) and xylose-lysine-deoxycholate (XLD) agars. In addition, an "enrichment" broth containing suppressants for normal enteric flora may be inoculated. After an incubation period to allow enteric pathogens to multiply, the enrichment broth is subcultured onto selective and differential agar plates to permit isolation of the pathogen from among the suppressed normal flora. Subsequent identification procedures are based on the same types of biochemical tests that you have studied, but may be more extensive to differentiate the enzymatic activities of enteric species that are closely related to *Salmonella* or *Shigella*.

Other bacterial pathogens that are not members of the *Enterobacteriaceae* are associated with intestinal disease. *Campylobacter jejuni,* a curved, gram-negative bacillus, is the most common bacterial agent of diarrhea in children and young adults (see **colorplate 6**). It has relatively strict growth requirements, and special procedures must be used to isolate it in the laboratory. Some vibrios, notably *Vibrio cholerae* (the agent of cholera) and *Vibrio parahaemolyticus* (of the family

Table 23.1 Enterobacteriaceae

Important Genera	Pathogenicity	Lactose	I	M	Vi	C	Motility	H$_2$S	Urea	PD	LD	OD
Salmonella	Typhoid fever Gastroenteritis	−	−	+	−	+	+	+	−	−	+	+†
Shigella	Bacillary dysentery	− or late	±*	+	−	−	−	−	−	−	−	±
Escherichia	Normal flora in GI tract Urinary tract infection Infant and traveler's diarrhea Hemorrhagic colitis	+	+	+	−	−	±	−	−	−	+	±
Citrobacter	Normal flora Urinary tract infection	+	−	+	−	+	±	±	−	−	−	±
Klebsiella	Respiratory infection Urinary tract infection	+	−	−	+	+	−	−	+ late	−	+	−
Enterobacter	Normal flora Urinary tract infection	+	−	−	+	+	+	−	− or late	−	+	+
Serratia	Normal flora Urinary tract infection Nosocomial infection	− or late	−	−	+	+	+	−	−	−	+	+
Proteus	Normal flora Urinary tract infection	−	±	+	±	±	+	±	+ rapid	+	−	±
Providencia	Normal flora Urinary tract infection	−	+	+	−	+	+	−	−	+	−	−
Yersinia	Gastroenteritis Mesenteric adenitis	−	±	+	−	−	+ (25°C) − (35°C)	−	+	−	−	+

*± = Some species or strains +, some −.
†Salmonella serotype Typhi is OD negative.

Vibrionaceae), represent other examples of significant intestinal tract pathogens. These organisms can also be isolated from cultures of fecal material and identified by their characteristic morphological and metabolic properties. Although the choice of isolation media and identification procedures must be varied according to the nature of the organism being sought in a specimen, the principles are the same as those we are following here. You should read further about infectious diseases acquired through the alimentary tract, including bacterial food poisonings, and be prepared to discuss the essential features of their laboratory diagnosis, beginning with the collection of appropriate specimens.

Conventional culture techniques and biochemical tube and immunoassay tests are gradually being replaced by molecular gene amplification methods that detect a large number of intestinal pathogens directly in stool specimens; for example, the multiplex PCR assay that is briefly reviewed in Exercise 17 (Principles of Nucleic Acid Assays and Multiplex Syndrome Panel Testing for the Diagnosis of Infectious Disease). One such multiplex assay can detect 20 common diarrheal disease pathogens, including bacteria, viruses, and protozoa. The test is more sensitive than conventional culture methods, and final results are available within 1 hour of specimen receipt instead of after several days when the specimen is processed by culture. The prompt diagnosis of an enteric infection in such a short time improves patient care and permits the appropriate administration of antibiotics, if indicated. In addition, if the patient is a food worker or employed in a hospital, nursing home, or day care center, infection control measures can be implemented several days sooner compared to culture.

In this experiment and Experiment 23.3, we shall review the basic methods for isolation and identification of enteric pathogens belonging to the genera *Salmonella* and *Shigella*. The general procedures are summarized in the flowchart shown in figure 23.1, and the biochemical reactions that you have studied in identifying *Enterobacteriaceae* are reviewed in table 23.1.

Purpose	To observe the morphology of *Salmonella* and *Shigella* species on selective and differential isolation plates
Materials	MacConkey plates
	Hektoen enteric (HE) plates
	Xylose-lysine-deoxycholate (XLD) agar plates
	Agar slant cultures of a *Salmonella* species and a *Shigella* species
	Wire inoculating loop
	Marking pen or pencil
	Test tube rack

Procedures

1. Inoculate a *Salmonella* culture on each of the properly labeled selective media provided. Streak for isolation. Do the same with a *Shigella* culture.
2. Incubate your six plates at 35°C for 24 hours.
3. Examine all plates and record your observations under Results.

Results

	Colonial Morphology on		
Name of Organism	*MacConkey*	*HE*	*XLD*
Salmonella			
Shigella			

Look up the composition of HE agar. List the major ingredients and state why you think they should affect the appearance of *Salmonella* in the way you have reported.

CASE STUDY

A Case of Diarrhea

A mother brought her 2-year-old son to the emergency room because he was experiencing a 103°F (39.4°C) fever, abdominal pain, and bloody diarrhea. The child's medical history was unremarkable, but his mother had taken him to a "fast-food" restaurant 2 days previously, where he ate a hamburger that appeared "red on the inside with bloody juices." The microorganism responsible for the boy's illness was isolated from a stool sample submitted to the microbiology laboratory for testing. On the basis of the patient's history:

1. Name the microorganism that is most likely responsible for this child's diarrhea.
2. Where is this organism normally found in nature?
3. How did the child most likely acquire this infection?
4. What laboratory tests are available to detect this microorganism in a stool sample?
5. Name a life-threatening complication of this disease that can develop in young children especially.
6. How can this disease be prevented?

EXPERIMENT 23.3   Identification Techniques for Enteric Pathogens

Purpose	To study some biochemical reactions of *Salmonella* and *Shigella*
Materials	TSI slants MR-VP broths Simmons citrate slants Urea broth tubes Phenylalanine agar slants Lysine and ornithine decarboxylase broths Mineral oil in dropper bottle Sterile 1.0-ml pipettes Pipette bulb or other aspiration device Sterile empty tubes Cotton-tipped swabs Kovac's reagent Methyl red indicator 5% alphanaphthol 40% potassium hydroxide 10% ferric chloride Agar slant cultures of *Salmonella* and *Shigella* species Wire inoculating loop Marking pen or pencil Test tube rack

Procedures

1. You will be assigned a culture of either *Salmonella* or *Shigella*. Label all tubes with the name of your organism. Inoculate one tube of each medium provided (i.e.: TSI; MR-VP broth; citrate slant; urea, lysine, and ornithine broths; and phenylalanine agar).
2. Incubate all tubes at 35°C for 24 hours.
3. Complete the IMViC and PD tests (see Experiment 23.1). Read and record all biochemical reactions under Results. Observe your neighbors' results and record all information for both organisms.

Results*

Name of Organism	TSI		I	M	Vi	C	H$_2$S	Urease	PD	LD	OD
	Slant	Butt									
Salmonella											
Shigella											

*Refer to Experiment 23.1 for explanation of tests and abbreviations.

EXPERIMENT 23.4   Serological Identification of Enteric Organisms

In addition to culture identification techniques, antibody reagents are available to detect O and H antigens of gram-negative enteric bacilli (usually *Salmonella* serotypes, *Shigella* species, and *Escherichia coli*). The antibodies are used in a simple bacterial agglutination test in which an unknown organism isolated in culture is mixed with the antibody reagent (antiserum). If the antibodies are specific for the organism's antigenic makeup, agglutination (clumping) of the bacteria occurs. If the antiserum does not contain specific antibodies, no clumping is seen. A control test in which saline is substituted for the antiserum must always be included to be certain that the organism does not clump in the absence of the antibodies.

Diagnostic Microbiology in Action

In this experiment, you will see how a microorganism can be identified by an interaction of its surface antigens with a known antibody that produces a visible agglutination of the bacterial cells.

Purpose	To illustrate identification of a microorganism by the slide agglutination technique
Materials	Glass slides 70% alcohol Saline (0.85%) Capillary pipettes Heat-killed suspension of *E. coli* or *Salmonella* *E. coli* or *Salmonella* antiserum Marking pen or pencil

Procedures

1. Carefully wash a slide in 70% alcohol and let it air dry.
2. Using a glass-marking pen or pencil, draw two circles at opposite ends of the slide.
3. Using a capillary pipette, place a drop of saline in one circle. Mark this circle "C," for control.
4. With a fresh capillary pipette, place a drop of antiserum in the other circle.
5. Use another pipette to add a drop of heat-killed bacterial suspension (this is the antigen) to the material in each circle.
6. Pick up the slide by its edges, with your thumb and forefinger, and rock it gently back and forth for a few seconds.
7. Hold the slide over a good light and observe closely for any change in the appearance of the suspension in the two circles.

Results

1. In the following diagram indicate any visible difference you observed in the suspensions at each end of the slide.

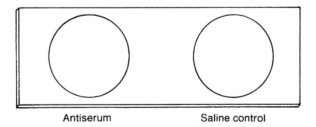

Antiserum   Saline control

2. State your interpretation of the result.

EXPERIMENT **23.5**   **Techniques to Distinguish Nonfermentative Gram-Negative Bacilli from *Enterobacteriaceae***

A variety of gram-negative bacilli that normally inhabit soil and water or live as commensals on human mucous membranes may contaminate specimens sent to the microbiology laboratory for culture or, more importantly, may produce opportunistic human infections. Typically, an opportunistic infection is one that occurs in an immunocompromised person, that is, one with lowered host defenses. The immunocompromising condition can result from underlying diseases, such as diabetes, or from the administration of steroids or chemotherapeutic agents. Opportunistic infections may occur also in people with burns to the skin or those who have major wounds to the skin or mucous membranes, because these protective barriers that normally exclude bacterial entry have been breached. Microorganisms that produce infections in such compromised patients are generally of low virulence and cause infections only when the "opportunity" is presented. Thus, they are called opportunistic pathogens, and the term opportunistic infection is used for the disease they produce in the host. Although the Gram-stain appearance and cultural characteristics of the organisms may resemble those of *Enterobacteriaceae*, they are relatively inactive in the common biochemical tests. In particular, they either fail to metabolize glucose or they degrade it by oxidative rather than fermentative pathways. For this reason, these organisms are often referred to as "glucose nonfermenters" (as opposed to the glucose-fermenting enteric bacilli). A number of bacterial genera and species are included in this group of nonfermenters. The most important from a medical aspect is *Pseudomonas aeruginosa,* which is most often involved in human infection. Because of the different clinical implications and the varying antimicrobial susceptibility patterns (nonfermenters are more highly resistant to common antimicrobial agents), it is important to distinguish nonfermenters from enteric bacilli. The characteristics of a few nonfermenting bacteria are listed in table 23.2 and compared with those of the *Enterobacteriaceae*.

Table 23.2   Characteristics of Nonfermenting Gram-Negative Bacilli

| | Butt of TSI | O-F glucose* | | Oxidase | Complete Hemolysis† | Diffusible Green Pigment |
		Open	Closed			
Pseudomonas aeruginosa	No change	+	–	+	+	+
Acinetobacter baumannii	No change	+	–	–	–	–
Acinetobacter lwoffi	No change	–	–	–	–	–
Alcaligenes faecalis	No change	–	–	+	–	–
Enterobacteriaceae	Yellow	+	+	–	– or +	–

*A positive test is a yellow color. Yellow in the open tube only indicates glucose degradation or *oxidation*. A yellow color in the closed tube (with mineral oil) indicates the organism is *fermentative* rather than oxidative. Glucose fermenters produce acid (yellow color) in the open as well as the closed tube.
†Around colonies on blood agar plates.

Purpose	To study some biochemical reactions of glucose nonfermenting bacteria
Materials	Blood agar plates
	Nutrient agar plates
	TSI slant
	O-F glucose deeps
	Oxidase reagent (di- or tetramethyl-*p*-phenylenediamine) or commercially available oxidase strips or disks
	Dropper bottle with sterile mineral oil
	Slant cultures of *Pseudomonas aeruginosa, Acinetobacter baumannii,* and *Escherichia coli*
	Wire inoculating loop
	Marking pen or pencil
	Test tube rack
	Sterile Petri dish
	Filter paper

Procedures

1. Label all plates and tubes as appropriate. Prepare and examine a Gram-stained smear of each organism.
2. Inoculate a blood and nutrient agar plate with each organism. Streak the plate to obtain isolated colonies.
3. Inoculate each organism onto a TSI slant by stabbing the butt and streaking the slant.
4. Inoculate *two* tubes of O-F glucose with each organism by stabbing your inoculating loop to the bottom of the column of medium. Overlay *one* of each set of two tubes with a one-half-inch layer of sterile mineral oil.
5. Label all plates and tubes. Incubate them at 35°C for 24 hours.
6. Test each organism for the presence of the enzyme *oxidase* by the following procedure or use the commercially available disks or strips. See Experiment 15.3, for a description of this test, and **colorplate 13**.
   a. Take a sterile Petri dish containing a piece of filter paper.
   b. Wet the paper with oxidase reagent.
   c. With your inoculating loop, scrape up some growth from the tube labeled *P. aeruginosa* and rub it on a small area of the wet filter paper. You should see an immediate *positive* oxidase reaction as the color of the area changes from light pink to black-purple.
   d. Repeat procedure 6c using growth from the tubes labeled *A. baumannii* and *E. coli*. Record the results in the following table.

Results

1. Examine the blood agar plate for hemolysis and the nutrient agar for pigment production.
2. Read and record all biochemical reactions in the following table.

Name of Organism	Gram-Stain Appearance		Butt of TSI (Color)	O-F Glucose		Oxidase (+ or −)	Complete Hemolysis (+ or −)	Diffusible Green Pigment (+ or −)
	Blood Agar	Nutrient Agar		Open (+ or −)	Closed (+ or −)			
Pseudomonas aeruginosa								
Acinetobacter baumannii								
Escherichia coli								

EXPERIMENT 23.6   Rapid Methods for Bacterial Identification

The biochemical tests performed in the preceding sections are representative of standard methods for bacterial identification. All biochemical testing must be performed with pure cultures to ensure the correct identification of the bacterial isolate. In some instances, it is possible to identify a bacterium correctly by using only a few tests, but more often an extensive biochemical "profile" is needed. Because it is expensive and time consuming to make and keep a wide variety of biochemical test media on hand, many microbiology laboratories now use multimedia identification kits. These are commercially available and are especially useful for identifying the common enteric bacteria. The use of such kits is customarily referred to as an application of "rapid methods," even though they must be incubated overnight, as usual, before results can be read. Some of them, indeed, are rapid to inoculate, while others permit complete identification within 24 hours.

One type of kit, the BBL Enterotube II (BD Microbiology Systems), is a tube of 12 compartmentalized, conventional agar media that can be inoculated rapidly from a single isolated colony on an agar plate (see **colorplate 36**). The media provided indicate whether the organism ferments the carbohydrates glucose, lactose, adonitol, arabinose, sorbitol, and dulcitol; produces H2S and/or indole; produces acetylmethylcarbinol; deaminates phenylalanine (PAD); hydrolyzes urea; decarboxylates lysine and/or ornithine; and can use citrate when it is the sole source of carbon in the medium. The mechanism of the other tests provided by the Enterotube II has been described in previous exercises (14, 15) and experiment 23.1.

The API System (bioMérieux Inc.) represents another type of kit for rapid identification of bacteria. This system provides, in a single strip, a series of 20 microtubules (miniature test tubes) of dehydrated media that are rehydrated with a saline suspension of the bacterium to be identified (see **colorplate 36**). The tests included in the strip determine whether the organism ferments glucose, mannitol, inositol, sorbitol, rhamnose, saccharose, melibiose, and amygdalin; produces indole and H_2S; hydrolyzes urea; breaks down the amino acids tryptophan (same mechanism as phenylalanine), lysine, ornithine, and arginine; produces gelatinase; forms acetylmethylcarbinol from glucose (VP test); and splits the compound o-nitrophenyl-β-D-galactopyranoside (ONPG). The enzyme that acts on ONPG, called β-galactosidase, also is responsible for lactose fermentation. Some bacteria, however, are unable to transport lactose into their cells for breakdown, although they possess β-galactosidase. In lactose broth, therefore, such bacteria fail to display acid production, or do so only after a delay of days or weeks. By contrast, in ONPG medium their β-galactosidase splits the substrate in a matter of hours, producing a bright yellow end product. Thus, ONPG can be used for the rapid demonstration of an organism's ability to ferment lactose.

A third type of kit, MicroScan (Beckman Coulter), consists of a multiwell panel containing dried antimicrobial agents for susceptibility testing and biochemical reagents for identification of enteric and glucose nonfermenting gram-negative bacilli. The wells of the panel are inoculated with a standardized suspension of an organism, incubated for 16 to 24 hours at 35°C, and then read visually or in an automated instrument. In this way, antimicrobial susceptibility testing and organism identification are achieved simultaneously. For enteric organisms, the biochemicals present in the wells test for fermentation of carbohydrates (glucose, sucrose, sorbitol, raffinose, rhamnose, arabinose, inositol, adonitol, melibiose); production of urease, H_2S, and indole; breakdown of lysine, arginine, ornithine, tryptophan, and esculin; and VP and ONPG reactions. In addition, tests for glucose nonfermenters include O-F glucose; ability to grow on minimal media containing citrate, malonate, tartrate, and acetamide; and ability to reduce nitrate.

In order to permit more accurate bacterial identification, a numeric recognition system has been devised for each of these three kits that assigns a number to each positive biochemical reaction. These figures are grouped together to give a numerical code to each organism. Unknown bacteria can be identified by looking up the code number provided by their positive reactions in an index book. Different strains of the same bacterium may vary in certain biochemical test results and thus have different code numbers. These variations can sometimes be used as epidemiological markers.

A further advance is the use of *automated* instruments to read and interpret the results of both identification and antimicrobial susceptibility tests. The tests are set up in special, clear-plastic multiwelled chambers containing a battery of biochemicals and different concentrations of several antimicrobial agents. The plastic chambers are then incubated in the instrument, which periodically scans each biochemical well for changes in the color of pH indicators and scans the antimicrobial agent wells for the presence of turbidity (signifying resistance). At the end of a specific time period, the computer in the instrument interprets all reactions and then the organism identification and its antimicrobial susceptibility results are printed out. Depending on the system used, results can be obtained in as little as 2 to 6 hours.

In this experiment some rapid nonautomated methods for identification of bacteria will be demonstrated.

Purpose	To observe the biochemical properties of bacteria grown in a multimedia system for rapid identification
Materials	Two Enterotubes, API strips, or MicroScan panels (as available) inoculated, respectively, with *Escherichia coli* and *Proteus vulgaris,* and incubated for 24 hours. The instructor will demonstrate methods for completing each test in the system.

Results

1. If Enterotubes were inoculated, record the results observed for each organism in the blocks provided under the following diagram.

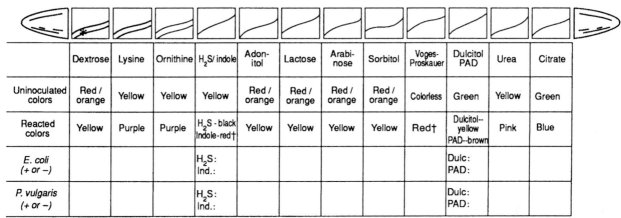

	Dextrose	Lysine	Ornithine	H₂S/indole	Adon-itol	Lactose	Arabi-nose	Sorbitol	Voges-Proskauer	Dulcitol PAD	Urea	Citrate
Uninoculated colors	Red / orange	Yellow	Yellow	Yellow	Red / orange	Red / orange	Red / orange	Red / orange	Colorless	Green	Yellow	Green
Reacted colors	Yellow	Purple	Purple	H₂S - black Indole-red†	Yellow	Yellow	Yellow	Yellow	Red†	Dulcitol-- yellow PAD--brown	Pink	Blue
E. coli (+ or −)				H₂S: Ind.:						Dulc: PAD:		
P. vulgaris (+ or −)				H₂S: Ind.:						Dulc: PAD:		

BD Diagnostic Systems.

*If this wax overlay is separated from the dextrose agar surface, gas has been produced by the organism.
†Requires addition of indole or Voges-Proskauer reagents.

2. If API strips were inoculated, record the results observed for each organism in the blocks provided under the following diagram.

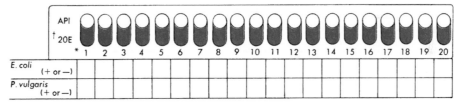

*Code	Test	Negative Reaction	Positive Reaction
1—ONPG	ONPG	Colorless	Yellow
2—ADH	Arginine dihydrolase	Yellow	Red or orange
3—LDC	Lysine decarboxylase	Yellow	Red or orange
4—ODC	Ornithine decarboxylase	Yellow	Red or orange
5—CIT	Citrate	Light green or yellow	Blue
6—H₂S	Hydrogen sulfide	No black deposit	Black deposit
7—URE	Urea	Yellow	Red or orange
8—TDA	Tryptophan deaminase	Yellow	Brown-red
9—IND	Indole	Yellow	Red-ring
10—VP	Voges-Proskauer	Colorless	Red within 10 min.
11—GEL	Gelatin	No black pigment diffusion	Black pigment diffusion
12—GLU	Glucose	Blue or blue-green	Yellow or gray
13—MAN	Mannitol	Blue or blue-green	Yellow
14—INO	Inositol	Blue or blue-green	Yellow
15—SOR	Sorbitol	Blue or blue-green	Yellow
16—RHA	Rhamnose	Blue or blue-green	Yellow
17—SAC	Saccharose	Blue or blue-green	Yellow
18—MEL	Melibiose	Blue or blue-green	Yellow
19—AMY	Amygdalin	Blue or blue-green	Yellow
20—ARA	Arabinose	Blue or blue-green	Yellow

† 20 E = 20-test strip for enteric bacteria.

3. If the MicroScan panel was used, complete the following table (note that tests included are for enterics only).

Well	Reagent Added	Positive Reaction	Negative Reaction	E. coli (+ or −)	P. vulgaris (+ or −)
GLU	None	Strong yellow only	Orange to red		
SUC					
SOR					
RAF					
RHA	None	Yellow to yellow/orange	Orange to red		
ARA					
INO					
ADO					
MEL					
URE	None	Magenta to pink	Yellow, orange, or light pink		
H₂S	None	Black precipitate or button	No blackening		
IND	1 drop Kovac's reagent	Pink to red	Pale yellow to orange		
LYS					
ARG	None	Purple to gray	Yellow		
ORN					
TDA	1 drop 10% ferric chloride	Brown (any shade)	Yellow to orange		
ESC	None	Light brown to black	Beige to colorless		
VP	1 drop 40% KOH, then 1 drop alphanaphthol; wait 20 min.	Red	Colorless		
ONPG	None	Yellow	Colorless		

Name _____ Class _____ Date _____

Questions

1. What does the term IMViC mean?

2. Why is the IMViC useful in identifying *Enterobacteriaceae?* Are further biochemical tests necessary for complete identification?

3. What diagnostic test differentiates *Proteus* and *Providencia* species from other *Enterobacteriaceae?*

4. How is *E. coli* distinguished from *P. vulgaris* on MacConkey agar? On a TSI slant?

5. Instead of TSI, why would a slant medium containing only dextrose and lactose (not sucrose) be preferable for detecting *Y. enterocolitica?* Name such a medium.

6. What procedures, other than biochemical, are used to identify microorganisms?

7. What is the purpose of the control test run in parallel with bacterial agglutination?

8. What is the value of serological identification of a microorganism as compared with culture identification?

9. How does a streptococcal latex agglutination test (see Experiment 20.1) differ from the bacterial agglutination test that you just performed? What information is derived from each?

10. Describe two mechanisms by which *E. coli* can produce disease.

11. What is meant by the term "enteric pathogen"?

12. Name a bacterial pathogen, other than one of the *Enterobacteriaceae,* that causes intestinal disease. Provide a flowchart indicating how you would make the laboratory diagnosis.

13. Name a rapid method for the identification of *Enterobacteriaceae,* and discuss its value in comparison with the standard methods you have used in Exercise 23.

14. Why is it important to differentiate glucose nonfermenters from *Enterobacteriaceae*?

15. Why must all bacterial identifications be performed with pure cultures?

Clinical Specimens from the Intestinal Tract

Learning Objectives

After completing this exercise, students should be able to:
1. Describe the enteric bacteria that are normal flora of the intestinal tract.
2. Explain why the preparation of a Gram-stained smear from a stool specimen is not recommended.
3. List the enteric pathogens that the laboratory wishes to isolate when culturing a stool sample.
4. Discuss why several selective and differential media are used for detecting possible enteric pathogens.
5. Differentiate the antibiotic susceptibility patterns of gram-positive and gram-negative bacteria.

This exercise provides you with an opportunity to apply your knowledge of the *Enterobacteriaceae* to making a laboratory diagnosis of an intestinal infection. In Experiment 24.1 you will be provided with a simulated sample of feces and observe the normal intestinal flora on primary isolation plates. In Experiment 24.2 you will be given a pure culture of one of the *Enterobacteriaceae* as an "unknown" to be identified. Experiment 24.3 is an antimicrobial susceptibility test of your pure unknown culture. Here you should observe the differences in response of gram-negative enteric bacilli, as compared with streptococci and staphylococci studied earlier, to the most clinically useful antimicrobial agents.

EXPERIMENT **24.1**   **Culturing a Fecal Sample**

Purpose	To study some enteric bacilli normally found in the bowel
Materials	A simulated stool specimen
	MacConkey agar plate
	Hektoen enteric (HE) agar plate
	Blood agar plate
	Wire or disposable inoculating loop
	Incubator
	Marking pen or pencil

Procedures

1. A simulated stool sample will be provided by your instructor.
2. Using a sterile inoculating loop, inoculate the fecal suspension on labeled blood agar, MacConkey agar, and Hektoen enteric (HE) agar plates. Streak for isolation, using the loop.
3. Incubate the plates at 35°C for 24 hours.

Results

1. Describe the appearance of growth on your plate cultures.

Plate	Relative Number of Colonies	Number of Colony Types	Color of Colonies	Type of Hemolysis
Blood agar				
MacConkey				×
HE				×

2. Interpret any difference in numbers of colonies on these plates.

3. Interpret the color of colonies on MacConkey agar.

4. Interpret the appearance of colonies on the HE plate.

5. Were any lactose-negative colonies present? If so, name the genera to which they might belong and indicate the key procedures that would identify each.

EXPERIMENT **24.2**   **Identification of an Unknown Enteric Organism**

Purpose	To use the techniques you have learned to identify an unknown pure culture
Materials	Same as in Experiments 23.2 and 23.3 except that your assigned culture is numbered, not labeled

Procedures

1. Prepare a Gram stain of your culture.
2. Inoculate the culture on MacConkey and HE. Label the plates and streak for isolation.
3. Label and inoculate all tubed media provided.
4. Incubate plates and tubes at 35°C for 24 hours.

Results

Read and record your results across one line of the following table. Also record all results obtained by your neighbors with different isolates. Refer to Experiment 23.1 for explanation of tests and abbreviations.

| Specimen Number | Gram-Stain Reaction and Morphology | Culture Medium (Growth + or −) | | TSI (Lactose + or −) | Biochemical Reactions (+ or −) | | | | | | | | | | Motility (+ or −) | Identification |
		MacConkey	HE		I	M	Vi	C	H₂S	Urea	PD	LD	OD		

EXPERIMENT **24.3** **Antimicrobial Susceptibility Test of an Enteric Organism**

Purpose	To determine the antimicrobial susceptibility pattern of a gram-negative enteric bacillus
Materials	Nutrient agar plates (Mueller-Hinton, if available)
	Antimicrobial disks
	McFarland No. 0.5 turbidity standard
	Sterile swabs
	Forceps
	Pure plate or slant culture of unknown from Experiment 24.2
	Tube of nutrient broth (5.0 ml)
	Marking pen or pencil

Procedures

1. Using a sterile swab or inoculating loop, take some of the growth of the pure culture of your unknown organism and emulsify it in 5.0 ml of nutrient broth to equal the turbidity of a McFarland 0.5 standard. Discard the swab into a disinfectant solution or biohazard bag.
2. Take another sterile swab, dip it in the broth suspension, drain off excess fluid against the inner wall of the tube.
3. Inoculate an agar plate as described in Experiment 12.1.
4. Follow steps 4 through 7 of Experiment 12.1.
5. Incubate the agar plate at 35°C for 24 hours.
6. Examine plates and record results for each antimicrobial disk as S (susceptible), I (intermediate), or R (resistant).
7. Compare results with those obtained for the organism you tested in Experiments 12.1 and 22.3.

Results

Record your findings.

Antimicrobial Agent	Organism in Exp. 12.1 Name:	S	I	R	Organism in Exp. 22.3 Name:	S	I	R	Organism in Exp. 24.3 Name:	S	I	R

1. Judging by the results of your tests, what group of antimicrobial agents appear to be indicated for the treatment of patients with gram-negative infections? Gram-positive infections?
2. What conclusions can you draw as to the importance of testing each suspected bacterial pathogen for its antimicrobial susceptibility?

Questions

1. What diseases are caused by *Salmonella*?

2. How do *Salmonella* enter the body? From what sources?

3. Name two selective media for the isolation of *Salmonella* and *Shigella*.

4. Name some of the normal flora of the intestinal tract.

5. Why is it not necessary to collect a stool specimen for culture in a sterile container?

6. How would you dispose of a real fecal specimen after inoculating cultures? How should cultures of feces be disposed of? Why?

7. Were the organisms in your simulated fecal culture predominantly lactose fermenters or nonfermenters? Does this have significance?

8. How do intestinal flora gain entry to the body?

9. Are the gram-negative enteric bacilli fastidious organisms? Would they survive well outside of the body? If so, what significance would this have in their transmission?

Urine Culture Techniques

Learning Objectives

After completing this exercise, students should be able to:

1. Describe three ways in which urinary tract infections may be acquired.
2. Explain the difference between significant and insignificant bacteriuria.
3. Define cystitis, pyelonephritis, and acute urethral syndrome.
4. Name three gram-negative and three gram-positive bacteria that may cause urinary tract infections.
5. Instruct male and female patients on the proper collection of a clean-catch, midstream urine sample.

Normally, urine is sterile when excreted by the kidneys and stored in the urinary bladder. In health, the only portion of the urinary tract that contains a normal flora is the terminal third of the urethra. When urine is voided, however, it becomes contaminated by the normal flora of the urethra and other superficial microorganisms found on urogenital mucous membranes. The presence of bacteria in voided urine (*bacteriuria*), therefore, does not always indicate urinary tract infection. To confirm infection, either the numbers of organisms present or the species isolated must be shown to be significant.

Active infection of the urinary tract develops in one of three ways: (1) microorganisms circulating in the bloodstream from another site of infection are deposited and multiply in the kidneys to produce *pyelonephritis* by the *hematogenous* (originating from the blood) *route*; (2) bacteria colonizing the terminal third of the urethra or the external urogenital surfaces ascend the urethra to the bladder, causing *cystitis* (infection of the bladder only) or pyelonephritis by the *ascending route*; or (3) microorganisms, usually from the urethra, are introduced into the bladder on catheters or cystoscopes.

Cystitis is much more common than pyelonephritis. In the former case, most of the offending organisms are opportunistic members of the fecal flora, including many of the gram-negative bacteria you have studied in Exercises 23 and 24, such as *Escherichia coli* (by far the most frequent cause of urinary tract infection). *Klebsiella, Enterobacter, Serratia,* and *Proteus, Pseudomonas,* and *Acinetobacter* are also incriminated, especially in hospitalized patients with indwelling urinary catheters or those receiving multiple antimicrobial agents. When these organisms reach the bladder, where active host-defense mechanisms (blood phagocytes and antibodies) are not readily available, they may grow in the urine, producing acute bladder infection. Among the gram-positive bacteria, *Enterococcus* spp., *Staphylococcus saprophyticus* and other coagulase-negative staphylococci, and *Staphylococcus aureus* may also cause urinary tract infections.

The blood that flows through the kidneys normally carries no microorganisms because phagocytic white blood cells and serum antibodies are constantly at work eliminating any microbial intruders that reach deep tissues. If these defense mechanisms are not working well or become overwhelmed by extensive infectious processes in systemic tissues (uncontrolled tuberculosis or yeast infections, staphylococcal or streptococcal abscesses), then the kidneys may become infected by organisms carried to them via the bloodstream. More commonly, however, pyelonephritis results from microorganisms initially infecting the bladder and then ascending the ureters to infect the kidneys.

Laboratory diagnosis of urinary tract infection is made by culturing urine, usually obtained either by catheterization or by voided collection. To obtain a catheterized urine specimen, a sterile, polyurethane catheter tube is inserted into the urethra and passed up into the bladder. The urine drains through the catheter tube and is collected in a sterile specimen cup. If it is obtained correctly, catheterized urine is not contaminated by normal urogenital flora and represents urine obtained directly from the bladder. The risk of catheterization, however, is that the technique may introduce microorganisms into the bladder that could result in infection, particularly if the patient does not already have cystitis. For this reason, voided urine specimens are more commonly collected for urine culture. In culturing voided specimens, however, the laboratory is faced with several problems. One is the normal contamination of voided urine by urogenital flora; another is the need for speed in initiating culture before contaminants can multiply and distort results; and a third is the obligation to obtain and report results that accurately reflect the clinical significances of the isolates. Contamination by extraneous organisms can mask the presence of other pathogens that are difficult to cultivate on artificial media. Overgrowths in urine stored at room temperature give a false picture of numbers. Either situation can lead to laboratory results that fail to reveal the clinical problem, and possibly to the mismanagement of the patient's case.

To address these problems, the laboratory must insist on adherence to proper techniques for urine collection and on prompt delivery of specimens to the laboratory for culture. When delay is unavoidable, urine specimens should be refrigerated to prevent multiplication of any microorganisms they may contain. Alternatively, a novel urine transport system that inhibits the growth of bacteria in urine without refrigeration has been developed. The system consists of a sterile evacuated tube that contains boric acid. Once the urine sample is collected, it is aspirated immediately into the evacuated tube. The boric acid, which is nontoxic to bacteria, disperses throughout the urine and inhibits bacterial growth in the sample for up to 12 hours at room temperature. Upon receipt in the laboratory, a *urinalysis* may be performed on the urine sample, which screens for certain physical and chemical properties that can indicate infection (i.e., turbidity, pH, protein, blood, and white blood cells). Uncontaminated urine is usually clear, but sometimes may be clouded with precipitating salts. Urine containing actively multiplying bacteria is turbid. If the patient has a urinary tract infection, the urine usually also contains many white blood cells. These white blood cells (leukocytes) may be observed microscopically or detected by performing a rapid dipstick test that screens for the presence of the enzyme leukocyte esterase. Leukocyte esterase is released when white blood cells die and, therefore, the detection of this enzyme in the urine sample is evidence that white blood cells are present. The rapid urine dipstick test is performed by dipping a chemically impregnated plastic strip into the urine sample and observing it for an appropriate color change. The same dipstick also tests for the presence of other chemicals in urine, such as protein and nitrate, which may also indicate infection when they are detected in abnormal amounts.

In some instances, the mere recovery of a pathogenic bacterial species in urine (e.g., *Salmonella, Mycobacterium tuberculosis,* or beta-hemolytic streptococci) is significant, regardless of numbers, and the search for such organisms does not require quantitative culture technique. It is generally advisable, however, to culture urine quantitatively, and to report a "colony count"—that is, the numbers of colonies that grow in culture from a measured quantity of urine. If microorganisms are actively infecting the kidneys or bladder, they can usually be demonstrated in large numbers in urine (in excess of 100,000 organisms per milliliter of urine). The recovery of greater than 100,000 bacteria per milliliter of urine in a properly collected and transported urine specimen is referred to as *significant bacteriuria* because the presence of such large numbers of bacteria in urine correlates with active infection of the bladder or kidney. On the other hand, normal urine that is merely contaminated in passage down the urethra contains very few organisms (100 to 1,000 per milliliter, not more than 10,000), *provided* it is cultured soon after collection, before multiplication of contaminants can occur in the voided specimen awaiting culture. The presence of less than 10,000 organisms per milliliter of urine in a voided specimen is sometimes referred to as *insignificant bacteriuria,* because this colony count does not represent infection. Some patients with symptoms of cystitis have low counts of the causative agent in their urine. This condition, called "acute urethral syndrome," is thought to occur in a subset of patients, usually women, who have symptoms of urinary tract infection. These patients require treatment, even when their urine cultures show low colony counts or no growth of bacteria. Close collaboration between the laboratory and the physician is needed to accurately diagnose these infections.

Collection of Voided Urine for Culture ("Clean-Catch" or "Clean-Voided" Techniques)

Aseptic urine collection requires careful cleansing of the external urogenital surfaces, using gauze sponges moistened with tap water and liquid soap. Special "clean-catch" urine collection kits are also available in many medical facilities. They contain moistened towelettes for proper cleansing of the genitalia and a sterile cup with a leak-proof, screw-cap lid for urine collection.

For males, the procedure simply entails thorough washing of the glans of the penis, discard of the first stream of urine, and collection of a "midstream" portion in a sterile container. If the outside of the container has been soiled with urine in the process, it must be wiped clean with disinfectant before being handled further.

For females, extra care is necessary. All labial surfaces must be thoroughly cleansed, and the sterile container must be held in such a way that it does not come in contact with the skin or clothing. Again, the first stream of urine is discarded, and a midstream sample is collected. When the container has been tightly closed, it is wiped clean with disinfectant. The proper procedures for the collection of "clean-voided, midstream" urine samples from male and female patients are shown in figure 25.1.

Figure 25.1   Procedure for collecting a clean-voided midstream urine sample.

The following steps will ensure that you obtain an accurate test:

1. Wash your hands with soap and water.

2. Remove the lid from the cup. Place it face up on the counter. Do not touch the inside of the cup or lid.

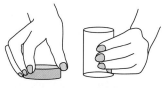

3. Wash your genitals using packaged towelettes or cotton balls with soap and water.

Women: Wipe between the folds of your genitals from front to back.

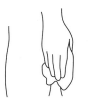

Men: Wipe the head of your penis and the opening. (Uncircumcised men pull foreskin back.)

4. Start to urinate into the toilet. Then, during midstream, move the cup into the stream to catch some urine. This will ensure that there are no bacteria in the urine sample.

5. Pull the cup away before you finish urinating when the cup is about half full. Do not overfill the cup.

6. Put the lid tightly on the cup and return it as directed. Be certain it is labeled appropriately.

Urine containers should never be filled to the brim. Closures should be double-checked to make certain they will not permit leakage during transport to the laboratory. If there is any delay (*before* or *after* delivery to the lab) in initiating culture, *urine specimens must be refrigerated.*

EXPERIMENT **25.1**   **Examination and Qualitative Culture of Voided Urine**

Purpose	To learn simple urine culture technique and to appreciate the value of clean-voided urine collection
Materials	Urine sample no. 1
	Urine sample no. 2
	Sterile empty test tubes
	Sterile 5.0-ml pipettes
	Pipette bulb or other aspiration device
	Litmus or pH papers
	Blood agar plates
	MacConkey plates
	Marking pen or pencil
	Test tube rack

Procedures

1. Your instructor will provide you with two simulated urine samples labeled numbers 1 and 2. Sample number 1 represents a urine sample collected without any special cleansing of the urogenital surfaces, whereas sample number 2 was collected using the appropriate "clean-voided" technique.
2. Place about 1.0 ml of each urine sample in small sterile test tubes labeled "no. 1" and "no. 2." Hold the tubes to the light and examine urine for color and turbidity. Test the pH of each sample with litmus or pH paper. Record your observations under Results.
3. Going back to the original urine container (the test tube sample is now contaminated by the pH test), pipette a large drop of the "clean-voided" specimen (no. 2) onto a blood agar plate near the edge, and another drop onto a MacConkey plate. Spread the drop a little with your loop and then streak for isolation. Label the plates "no. 2."
4. Repeat step 3 with the casual urine collection (no. 1), labeling the plates "no. 1."
5. Incubate all plates at 35°C for 24 hours.
6. Examine the incubated plates for amount of growth, types of colonies, and microscopic morphology of colony types. Record observations under Results.

Results

1. Macroscopic appearance of urine.

Specimen	Color	Turbidity	pH
Clean voided (no. 2)			
Casual (no. 1)			

2. Culture results.

Specimen	Blood Agar			MacConkey Agar		
	Amount of Growth	Types of Colonies	Gram Stain and Morphology	Amount of Growth	Types of Colonies	Gram Stain and Morphology
Clean voided (no. 2)						
Casual (no. 1)						

Interpret any differences you observed in the *amount* of growth recovered from the two specimens.

Interpret differences in the amount of growth on blood agar and MacConkey plates for each specimen.

Interpret differences in the nature of growth obtained on blood agar and MacConkey plates for each specimen.

Interpret any finding of "no growth."

EXPERIMENT **25.2**   **Quantitative Urine Culture**

To distinguish contamination of urine by normal urogenital flora from urinary tract infection caused by the same organisms, it is usually necessary to determine the numbers of organisms present per milliliter of specimen. In general, counts in excess of *100,000 organisms per milliliter* are considered to indicate *significant bacteriuria,* if the collection technique was adequate and there was no delay in culturing the specimen.

A quantitative culture is prepared by placing a measured volume of urine on an agar plate and counting the number of colonies that develop after incubation. A calibrated loop that delivers 0.01 ml of sample is used to inoculate the plate. The number of colonies that appear from this 1/100th-ml sample is multiplied by 100 to give the number per milliliter. For example, if 15 colonies are obtained from 0.01 ml, there are 15 × 100, or 1,500, organisms present in 1 ml (assuming each colony represents one organism). Some clinical microbiology laboratories use a 0.001-ml calibrated loop instead of a 0.01-ml loop for performing quantitative urine cultures. In this case, the number of colonies that grow is multiplied by 1,000 to give the total colony count per milliliter. For example, if 15 colonies are obtained from 0.001 ml of urine sample, there are 15 × 1,000 or 15,000 colonies per milliliter of urine.

In this experiment, you will have a simulated urine specimen from a suspected case of urinary tract infection submitted with a request for quantitative culture.

Purpose	To learn quantitative culture technique and to see the effects of delay in culturing a voided urine specimen
Materials	Nutrient agar plates
	Calibrated loop (0.01-ml delivery)
	Sterile 5-ml pipettes
	Pipette bulb or other aspiration device
	Sterile empty tubes
	Simulated "clean-voided" urine from a clinical case of urinary tract infection
	Test tube rack
	Marking pen or pencil

Procedures

1. With the calibrated loop, transfer 0.01 ml of the urine specimen to the center of a labeled nutrient agar plate and streak across the drop in several planes so that the specimen is distributed evenly across the plate.
2. Incubate the plate at 35°C for 24 hours.
3. Go back to the original urine specimen and measure about 2.0 ml into each of two sterile, empty test tubes. Label one "refrigerator" and place it in the refrigerator. Label the other tube "room temperature" and leave it at room temperature at your station. Place the original specimen in the incubator for 24 hours.
4. Read your plates, count the colonies on each, and report the numbers of organisms per milliliter present in the urine specimen.
5. Inspect the tubes of urine left in the refrigerator, in the incubator, and on your bench. Examine for turbidity and record results.

Results

1. Record the number of colonies on the streaked nutrient agar plate.

2. Calculate and record the number of organisms per milliliter of specimen and indicate whether this result is significant of urinary tract infection.

3. Diagram your observations of turbidity in each tube of stored urine.

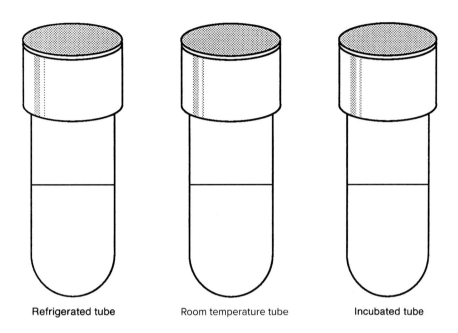

Refrigerated tube   Room temperature tube   Incubated tube

What is your interpretation of the appearance of these tubes?

CASE STUDY

A Case of Cystitis

A 21-year-old, sexually active female visited her family physician complaining of pain on urination (similar to a "burning sensation") and the production of foul-smelling urine. The physician suspected a urinary tract infection and ordered urinalysis and urine culture tests on a "clean-voided, midstream" urine sample. Urinalysis revealed the microscopic presence of greater than 40 white blood cells and 5 to 10 red blood cells per high-power field. In addition, on a dipstick test, the leukocyte esterase and nitrate reduction tests were both positive. A quantitative urine culture reported growth of greater than 100,000 gram-negative, lactose-fermenting bacilli per milliliter on MacConkey agar medium.

1. What is the significance of the urinalysis findings? Explain your answer.
2. Why do women usually experience more urinary tract infections than do men? Does being sexually active contribute to the problem?
3. What is the significance of the quantitative urine culture results?
4. Which bacterium is most likely responsible for this patient's infection?

Questions

1. What is bacteriuria? When is it significant?

2. How do microorganisms enter the urinary tract?

3. Why is aseptic urine collection important when cultures are ordered?

4. List five bacteria that can cause urinary tract infection.

5. If you counted 20 colonies from a 0.01-ml inoculum of a 1:10 dilution of urine, how many organisms per milliliter of specimen would you report? Is this number significant?

6. Is the urine colony count an appropriate indicator of the need for an antimicrobial susceptibility test of an organism isolated from a urine culture? Why?

7. If you took a urine specimen for culture to the laboratory but found it temporarily closed, what would you do?

8. How would you instruct a female patient to collect her own urine specimen by the "clean-catch" technique? A male patient?

9. What can you learn from visual inspection of a urine specimen?

10. Describe a urine transport system that allows the specimen to remain at room temperature for short time periods without refrigeration.

11. What is leukocyte esterase? What is its significance when detected in urine?

Neisseria and Spirochetes

Learning Objectives

After completing this exercise, students should be able to:

1. Provide examples of sexually transmitted diseases that are caused by a spirochete, an intracellular pathogen, and a gram-negative diplococcus.
2. Name the two species of pathogenic *Neisseria* and the diseases they cause.
3. Discuss the methods available for establishing the laboratory diagnosis of gonorrhea.
4. Know the screening and confirmatory tests that may be used to establish the serologic diagnosis of syphilis.
5. Name the organism that causes Lyme disease and discuss how the disease is acquired.

The sexually transmitted diseases are perhaps the most important infections acquired through the urogenital tract. Three common infectious diseases of this type are gonorrhea, syphilis, and chlamydial urethritis/cervicitis. All three infections are caused by bacteria. Gonorrhea is caused by *Neisseria gonorrhoeae;* syphilis by *Treponema pallidum,* a spirochete; and chlamydial infection by *Chlamydia trachomatis. Neisseria gonorrhoeae* can be grown on special laboratory culture media, but chlamydiae are obligate intracellular parasites (once considered viruses, in part for this reason) and require special laboratory techniques for isolation (see Exercise 30). *Treponema pallidum,* on the other hand, has not yet been grown in any laboratory culture system and is cultivated only in certain animals, such as the rabbit.

The bacterial groups to which these sexually transmitted agents belong contain other pathogenic species associated with nonsexually transmitted disease; that is, infections acquired through other entry portals. Lyme disease is one example of a nonsexually transmitted spirochetal disease that is transmitted to humans following the bite of an infected tick. The spirochete responsible for Lyme disease is *Borrelia burgdorferi* (see fig. 26.3). Still other species of *Neisseria* and *Treponema* are nonpathogenic, including some that are frequent members of the normal flora of various mucous membrane surfaces, particularly of the respiratory tract.

EXPERIMENT **26.1** *Neisseria*

The genus *Neisseria* contains two pathogenic species and a number of others that are commonly found in the normal flora of the upper respiratory tract. The two medically important species are *N. gonorrhoeae,* the agent of gonorrhea, and *N. meningitidis,* an agent of bacterial meningitis. All *Neisseria* are gram-negative diplococci (see **colorplate 5**), indistinguishable from each other in microscopic morphology. The pathogenic species are obligate human parasites and quite fastidious in their growth requirements on artificial media. Although they are aerobes, on primary isolation the pathogens require an increased level of CO_2 during incubation at 35°C. The nonpathogenic commensals of the upper respiratory tract are not fastidious and grow readily on simple nutrient media. Some of the respiratory flora, for example, *N. subflava* and *N. flavescens,* have a yellow pigment, but most *Neisseria* produce colorless colonies. All *Neisseria* are oxidase positive (see **colorplate 13**), which helps to distinguish them from other genera, but not from each other. Biochemically, the *Neisseria* species are most readily identified on the basis of their differing patterns of carbohydrate degradation. The cultural differentiation of a few *Neisseria* species, including the two pathogenic species, is shown in table 26.1. Nucleic acid probe tests and gene amplification assays for detecting *N. gonorrhoeae* directly in patient urogenital specimens have become available and are widely used for detecting *N. gonorrhoeae* in urogenital specimens (see Exercise 17). These tests can be completed within 2 to 4 hours and thus, diagnostic results can be available the day the specimen is taken and received by the laboratory.

Table 26.1 Differentiation of Some *Neisseria* species

Name of Organism	Pathogenicity	Growth on Enriched Media (in CO_2)	Growth on Simple Nutrients (in Air)	Yellow Pigment	Oxidase	Acid Production from*		
						G	M	S
N. gonorrhoeae	Gonorrhea	+	–	–	+	+	–	–
N. meningitidis	Meningitis	+	–	–	+	+	+	–
N. sicca	Normal flora respiratory tract	+	+	±	+	+	+	+
N. flavescens	Normal flora respiratory tract	+	+	+	+	–	–	–

*G = glucose; M = maltose; S = sucrose.

Figure 26.1 Diagram of a microscopic field showing intracellular diplococci within polymorphonuclear cells. In cervical smears, organisms of the normal flora may be numerous, but these are extracellular.

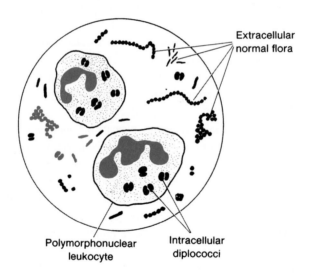

Even though clinical microbiology laboratories commonly use direct nucleic acid probe and gene amplification assays to detect urogenital infection, in some states these methods cannot be used for medical–legal cases, such as sexual abuse of children. The reason is that even though the molecular-based assays are now regarded as more sensitive than culture for detecting infection, there is a very small risk that sample contamination could result in a false-positive test report and result in the conviction of an innocent person. In such instances, the cultural isolation and confirmatory identification of the organism are the only test results admissible in court. In addition, because of the potential medical–legal ramifications of alleged child sexual abuse, strict protocols regarding the collection, transport, and laboratory testing of specimens must be followed. These protocols are referred to as *chain of custody.* Adherence to the chain of custody protocol ensures that the defendant's sample is not inadvertently mixed up with a specimen collected from a different individual.

Gonorrhea usually begins as an acute, localized infection of the genital tract. In the male, the urethra is initially involved and exudes a purulent discharge. When the exudate is smeared on a microscope slide and Gram stained, it is seen to contain many polymorphonuclear cells (phagocytic white blood cells), some of which contain intracellular, gram-negative diplococci. In the female, acute infection usually begins in the cervix. Smears of the exudate show the same intracellular diplococci as seen in males, except that there are often many more extraneous organisms present in specimens collected from females (see fig. 26.1 and

Figure 26.2 Recommended procedures for laboratory diagnosis of gonorrhea by smear and culture.
Source: Centers for Disease Control and Prevention, U.S. Public Health Service, Atlanta, Georgia, modified.

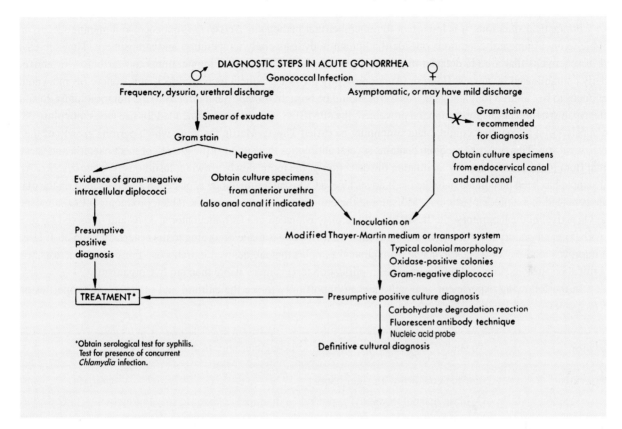

colorplate 5). Indeed, the abundant normal flora of the vagina may mask the presence of gonococci (*N. gonorrhoeae*) in smears or cultures from females. For many women, initial infection may be asymptomatic. In them, gonococci cannot be demonstrated in smears, and, therefore, culture or other detection techniques, such as nucleic acid probes or gene amplification assays, *must* be used for laboratory diagnosis. Asymptomatic infection also occurs in a smaller percentage of infected males. Demonstration of *N. gonorrhoeae* in culture or by a probe or gene amplification test provides definitive proof of infection. For culture, specimens may be taken from the cervix, urethra, anal canal, or throat, and one or more of these sites should be swabbed for culture when disease is suspected in women. For male patients with a urethral discharge, Gram-stained smears of the exudate revealing typical gram-negative intracellular diplococci are considered presumptively diagnostic, and cultures are generally not taken. Figure 26.2 outlines the recommendations of the Centers for Disease Control and Prevention, U.S. Public Health Service, regarding smears and cultures from males and females, indicating all necessary steps to confirm the diagnosis of gonorrhea by these methods.

When swab cultures are taken, a suitable agar medium should be inoculated directly with the swab, the culture placed in a candle jar or CO_2 incubator, and incubated at 35°C pending laboratory examination. Media enriched with hemoglobin and other growth factors are in common use (modified Thayer-Martin medium, for example). Antimicrobial agents are added to suppress the normal flora of mucous membranes and to make these media more selective for gonococci. Following suitable incubation, laboratory identification of *N. gonorrhoeae* is made by the criteria shown in table 26.1 (see colorplate 37).

Meningitis, an inflammation of the meninges of the brain, may be caused by a variety of microbial agents. Chief among them is *Neisseria meningitidis,* a gram-negative diplococcus. The usual portal of entry for these organisms is the upper respiratory tract. They may colonize harmlessly there in the immune individual. When they enter susceptible hosts who cannot keep them localized, they may cause invasive disease, either by finding their way into the bloodstream and then to deep tissues, or by direct extension through the membranous bony structures posterior to the pharynx and sinuses. When they localize on the meninges (the thin membranes that cover the brain), meningococci (*N. meningitidis*) induce an acute, purulent local infection that may have far-reaching effects in the central nervous system. The laboratory diagnosis of *N. meningitidis* infections is made by recovering the organism in cultures of spinal fluid or blood and identifying it by the criteria indicated in table 26.1. Fortunately, the incidence of serious meningococcal disease has been significantly reduced by the widespread use of effective vaccines that protect against infection caused by the most common serogroups of *N. meningitidis*.

In practical situations, it is important to remember that pathogenic *Neisseria* (gonococci and meningococci) are very susceptible to environmental conditions outside the human body, especially temperature and atmosphere. They are easily destroyed in specimens that are (1) delayed in transit to the laboratory, (2) kept at temperatures too far below or above 35°C, (3) heavily contaminated by normal flora, or (4) not promptly provided with an increased CO_2 atmosphere (as in a candle jar). All specimens to be cultured for pathogenic *Neisseria* should be brought *promptly* and *directly* to the microbiologist. In situations when delays in specimen transport cannot be prevented, the JEMBEC system is recommended for use (see **colorplate 38**). This system consists of a rectangular culture plate containing modified Thayer-Martin medium, which permits growth of the pathogenic *Neisseria* spp. This selective medium contains several antibiotics that suppress the growth of most bacteria and yeast found as normal flora in the vagina, thereby facilitating the detection of the more slowly growing, fastidious *N. gonorrhoeae*. After the clinical sample has been inoculated onto the medium surface, a CO_2-generating tablet is placed in a well located on the plate, and the plate is placed in a ziplock plastic bag. Moisture in the culture medium activates the tablet, producing a CO_2 atmosphere in the bag. On arrival in the laboratory, the JEMBEC culture plate is placed in a CO_2 incubator at 35°C and observed for growth as usual. As before, laboratory identification of suspected *N. gonorrhoeae* is made according to the criteria in table 26.1. Use of this system improves recovery of *N. gonorrhoeae* from clinical samples that are delayed in transport. For nonculture tests, transport is less critical because nucleic acid (probe or gene amplification test) is more stable than are live organisms.

In the following experiment, you will have an opportunity to see the cultural and microscopic properties of some *Neisseria* species.

Purpose	To study the microscopic and cultural characteristics of *Neisseria* species
Materials	Sterile Petri dish with filter paper
	Oxidase reagent (dimethyl-*p*-phenylenediamine) or oxidase strips or disks
	Phenol red broths or cystine trypticase agar (CTA) slants, each containing glucose, maltose, or sucrose
	Candle jar, containing preincubated chocolate agar plate cultures of:
	Neisseria sicca (pure culture)
	Neisseria flavescens (pure culture)
	A cervical exudate (simulated clinical specimen from a female giving a history of contact with a positive male patient with gonorrhea)
	Wire inoculating loop
	Marking pen or pencil

Procedures

1. Examine the morphology of colonies on each plate and record their appearance, including pigmentation.
2. Test representative colonies on *each* pure culture plate for the enzyme *oxidase* following the procedure in Experiment 23.5, step 6 (a–c).
3. Test representative colonies from the plated clinical specimen for their oxidase reactions. Using a marking pen or pencil, mark the bottom of the plate under colonies that are oxidase positive. Record oxidase reactions in the table under Results.
4. Make a Gram stain of an oxidase-positive colony from each of the chocolate plates. Record the microscopic morphology of each in the table under Results.
5. Inoculate one oxidase-positive colony from *each* chocolate agar plate into a glucose, maltose, and sucrose broth or CTA slant. Be certain to label the tubes appropriately.
6. Incubate all carbohydrate subcultures in a candle jar or CO_2 incubator at 35°C for 24 hours. Examine for evidence of acid production. Record results.

Results

1. Record your observations in the table that follows.

Culture	Colonial Morphology	Oxidase (+ or −)	Gram-Stain Morphology	Acid Production*		
				G	M	S
N. sicca						
N. flavescens						
Cervical specimen						
N. gonorrhoeae[†]						
N. meningitidis[†]						

*G = glucose; M = maltose; S = sucrose.
[†]To be completed from your reading.

2. Laboratory report of clinical specimen: _____

EXPERIMENT 26.2   Spirochetes

The spirochetes are slender, coiled organisms with a longitudinal axial filament that gives them motility. Seen in action, they are long, flexible, and always actively spinning or undulating. Their cell walls are extremely thin and not readily stainable. In unstained wet mounts they are too transparent to be seen by direct condenser light but become quite visible by "dark-field" condenser illumination, simply called *dark-field microscopy*. A special condenser lens is used to block the passage of direct light through the mount and to permit only the most oblique rays to enter, at an angle that is nearly parallel to the slide. When viewed in such minimal light, the background of the mount is very dark, almost black, but any particles in suspension are brightly illuminated because they catch and reflect light upward through the objective lens. To stain spirochetes in fixed smears, stains containing metallic precipitates are used. Silver, for example, can be precipitated out of solution and will coat spirochetes on the slide, giving them a black color when viewed by ordinary light microscopy (see **colorplate 7**).

There are three major genera of spirochetes: *Treponema, Borrelia,* and *Leptospira* (fig. 26.3), each containing species associated with human disease. Many of these organisms are obligate parasites that grow only in human or animal hosts, and others are difficult to cultivate on artificial culture media. The leptospires are an exception, for they will grow in a special serum-enriched medium or in embryonated eggs. The laboratory diagnosis of spirochetal diseases is made by microscopic demonstration of the organisms in appropriate clinical specimens, when possible; in special cultures in the case of leptospirosis; or, most frequently, by serological methods for detecting antibodies in the patient's serum (see Exercise 33).

Figure 26.3 Representative spriochetes. (a) *Treponema pallidum*; (b) *Borrelia* spp. resembling *Borrelia burgdorferi*; (c) *Leptospira interrogans*. Photos are for morphology only and not to scale.
(a) *Source: CDC/Schwartz*; (b) *©F1online digitale Bildagentur GmbH/Alamy Stock Photo*; (c) *©Science Source*

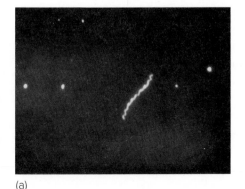

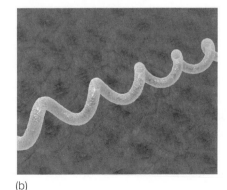

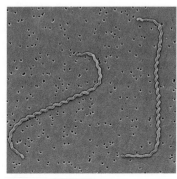

(a)   (b)   (c)

Treponema. The most important member of this genus is *Treponema pallidum,* the agent of syphilis. The organism can be demonstrated by dark-field examination (fig. 26.3a) of material from the primary lesion of the disease, called a *chancre.* Diagnostic serological tests for syphilis are numerous but can be divided into two major groups—*screening tests* and *confirmatory tests.* Screening tests examine patient serum for the presence of nonspecific antibodies as a possible indication that the patient *might* have syphilis. These screening tests are easy and inexpensive to perform, and allow for a large number of samples to be tested in a short time period. The Rapid Plasma Reagin (RPR) and the Venereal Disease Research Laboratory (VDRL) tests are two examples of serological tests that screen for nonspecific antibodies to syphilis. However, a positive RPR or VDRL test does not always mean that the patient has syphilis. The confirmatory test must be performed on the serum to determine if specific antibodies to *T. pallidum* are present in the sample. The confirmatory test is more time consuming and expensive to perform, but the detection of specific antibodies confirms that the patient has syphilis. The Fluorescent Treponemal Antibody Absorption test, often referred to as the FTA-ABS test, is an example of a confirmatory test for syphilis. More recently, many laboratories have replaced these costly, labor-intensive, screening and confirmatory assays with a single, more-sensitive enzyme immunoassay. The enzyme immunoassay specifically detects the presence of two types of treponemal antibodies, immunoglobulin M, or IgM, and immunoglobulin G, or IgG. The presence of IgM treponemal antibodies provides evidence that the patient has active, acute infection, whereas the presence of only IgG antibodies indicates that the patient has chronic, latent, or resolved infection. The absence of both antibodies indicates that the patient does not have syphilis. These serological tests are particularly valuable because syphilis can be a latent, silent infection, with few or no obvious symptoms in its early stages. It is a chronic, progressive disease, however, and if unrecognized and untreated, it can have very serious consequences. Also, it is a sexually transmitted disease, highly communicable in its primary stage. Laboratory diagnosis of syphilis is, therefore, essential in its recognition, treatment, and control.

Nonpathogenic species of *Treponema* are frequent members of the normal flora of the mouth and gums, and sometimes of the genital mucous membranes.

Borrelia. *Borrelia* species are pathogenic for humans and a wide variety of animals including rodents, birds, and cattle. They are transmitted by the bites of arthropods. Two important species for humans are *Borrelia recurrentis,* the agent of relapsing fever, and *Borrelia burgdorferi,* the agent of Lyme disease (fig. 26.3b).

Relapsing fever is now primarily a tropical disease, and is transmitted by lice. As the name implies, the infection is characterized by repeated episodes of fever with afebrile intervals in between. Diagnosis is made primarily by seeing the organisms in the patient's blood either in an unstained preparation viewed by dark-field microscopy or in a smear stained with routine dyes used in the hematology laboratory (e.g., Giemsa stain).

Lyme disease, transmitted by tick bites, occurs in a number of countries. In the United States, it was first recognized in children living in Lyme, Connecticut. Although once thought to be confined to the eastern part of the United States, this disease is a growing problem in many parts of the country and throughout the world. The first sign of infection is a circular, rashlike lesion resembling a bull's-eye (called erythema migrans) that begins at the site of the tick bite (see **colorplate 39**). However, not all infected patients develop the bull's-eye rash. This lesion may remain localized or spread to other body areas. The rash may be accompanied by flulike or meningitis-like symptoms, and if untreated, many patients develop arthritis, chronic skin lesions, and nervous system abnormalities after many weeks or even years. Because the signs and symptoms mimic those of other infections, the correct diagnosis is often not suspected. Currently, diagnosis is best made by detecting specific antibodies against the spirochete in the patient's serum, but a history of tick bite provides an important clue.

Leptospira. This genus had been classified as having only one species, *Leptospira interrogans,* but molecular studies show several species are included in this group. Several different serological strains are pathogenic for animals (dogs, rodents) and one is associated with a human disease called *icterohemorrhagia,* or, more simply, leptospirosis. The spirochetes infect the liver and kidney, producing local hemorrhage and jaundice, hence the clinical term icterohemorrhagia. Because *Leptospira* can infect the kidneys of dogs, rodents, and other mammals, large numbers of organisms may be excreted in the urine of infected animals. Individuals who come in direct contact with infected urine may develop leptospirosis from such exposure. The organism can also cause meningitis.

The laboratory diagnosis of leptospirosis can sometimes be made by demonstrating the organism in dark-field preparations of blood or urine specimens (this spirochete is very tightly coiled, its ends are characteristically hooked (fig. 26.3c), and it has a rapid, lashing motility). Culture of such specimens in serum medium (Fletcher's) or animal inoculation can also lead to recovery of the spirochete. Serological diagnosis can be made by testing the patient's serum for leptospiral antibodies.

In this experiment you will see the morphology of some spirochetes in prepared slides and demonstration material.

Purpose	Demonstration of important spirochetes
Materials	Prepared stained slides
	Projection slides, if available

Procedures

Examine the prepared material. From your observations and/or reading, illustrate the microscopic morphology of each spirochetal genus. Do they resemble the spirochetal forms in fig. 26.3?

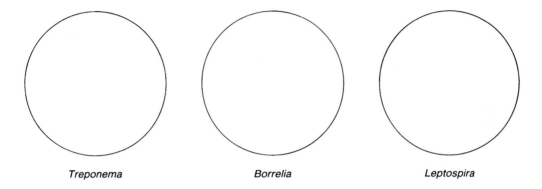

Treponema   Borrelia   Leptospira

CASE STUDY

A Case of a Patient with a Bull's-Eye Skin Lesion

An 8-year-old boy was in his usual state of good health after returning home with his family from a 10-day camping trip at a state park located on Long Island, New York. During their vacation, the family members wore short pants while hiking through the woods and fields. Within 2 days of returning home, the boy developed a low-grade fever and general fatigue. He had no symptoms other than a large, circular rash, resembling a bull's-eye, on his thigh.

1. Name the microorganism that is responsible for this boy's infection.
2. What is the name of the disease produced by this microorganism?
3. How did the boy likely acquire this infection?
4. What is the medical term for the bull's-eye lesion that developed on the patient's thigh?
5. What laboratory tests are available to establish the diagnosis of this disease?
6. The highest prevalence of this infection is in which geographic parts of the United States?
7. How can this infection be prevented?

Questions

1. Can you distinguish between *N. gonorrhoeae* and *N. meningitidis* by Gram stain? Explain.

2. What are intracellular gram-negative diplococci?

3. Why are selective media used for primary culture of specimens from the female urogenital tract?

4. Why is culture the standard method for diagnosing gonorrhea in possible medical–legal cases?

5. How are pathogenic *Neisseria* identified?

6. Name two common causes of bacterial meningitis.

7. When spinal fluid is collected for laboratory diagnosis of meningitis, how should it be transported to the laboratory?

8. Where are *Neisseria* found as normal flora? *Treponema*?

9. Name the etiologic agents of syphilis, leptospirosis, and Lyme disease.

10. Can *T. pallidum* be demonstrated by Gram stain? If not, what technique would you use to view this organism microscopically?

11. What is the importance of the laboratory diagnosis of syphilis and gonorrhea?

12. Provide a brief description of the screening and confirmatory tests for the serological diagnosis of syphilis.

13. What are the advantages of using the new enzyme immunoassay for establishing the serological diagnosis of syphilis?

14. Complete the following table.

Organism	Disease	Gram-Stain Reaction	Microscopic Morphology	Laboratory Diagnosis	
				Specimens	Method
Neisseria gonorrhoeae					
Neisseria meningitidis					
Treponema pallidum					
Borrelia burgdorferi					
Leptospira interrogans					

Clinical Specimens from Blood

Learning Objectives

After completing this exercise, students should be able to:

1. Define septicemia and bacteremia.
2. Explain the differences between the three major types of bacteremia.
3. Discuss the importance of skin antisepsis when collecting a blood specimen for culture.
4. Specify the recommended volumes of blood that must be collected when culturing blood from adult and pediatric patients.
5. Explain why collection of more than one blood culture specimen is required for establishing the diagnosis of bacteremia.

Microorganisms sometimes gain access to body sites that are normally sterile. Once there, they have the potential to cause serious, life-threatening disease. If they enter the central nervous system, they may cause meningitis or encephalitis; in joint fluid, they may cause septic arthritis. Those microorganisms that enter the bloodstream cause circulatory system infection, referred to as bacteremia, sepsis, or septicemia.

Septicemia is a serious condition in which microorganisms are present and persistent in the blood. The suffix -emia, which means "presence in the blood," is applied broadly to characterize microorganisms responsible for bloodstream infections. Students should become familiar with the term because its prefix characterizes the microbial etiology of infection. For example, bacteremia is the presence of bacteria in the blood, viremia refers to viruses in the blood, fungemia is used for fungal blood infections, and parasitemia refers to parasites in the blood. More specific terms refer to the actual causative organisms, for example, candidemia, spirochetemia, meningococcemia, and gonococcemia.

The purpose of this exercise is to familiarize students with the infections of the cardiovascular system and the laboratory methods available for their diagnosis.

Types of Bloodstream Infection

Although the terms bacteremia and septicemia are often used interchangeably, septicemia is a potential medical emergency because organisms persist in the bloodstream in large numbers. The disease has a high mortality rate if it is not promptly diagnosed and treated with the appropriate antibiotic(s). Although both septicemia and bacteremia can be caused by any microorganism, bacteremia is more common and is the focus of the discussion for the remainder of this exercise.

There are three different types of bacteremia: transient, intermittent, and continuous. *Transient bacteremia* occurs every day in healthy people and results from the occasional release of microorganisms into the bloodstream from a body site that contains normal flora. The microorganisms that enter the bloodstream are readily removed by the normal host defense mechanisms, such as phagocytosis by neutrophils and macrophages in conjunction with antibodies in the blood. Transient bloodstream infections produce no symptoms of infection (e.g., fever and shaking chills) and are often referred to as subclinical or inapparent infections. Normal daily activities, including flossing and brushing of teeth and a forced bowel movement, may cause a shower of normal flora organisms to enter the bloodstream from the relevant anatomic site. In most individuals, the microorganisms are readily removed by normal host defense mechanisms and no symptoms develop. In special cases, however, in persons who have heart valve disease, due to a congenital defect or rheumatic heart disease, or a heart valve replacement, an infection of the heart valve, referred to as *endocarditis,* may develop after an episode of transient bacteremia. In these patients, the circulating organisms become trapped and grow on the damaged or abnormal valve. For this reason, individuals with such heart valve conditions receive prophylactic antibiotics before dental work, including routine teeth cleaning, and when they undergo manipulation at body sites that have a rich normal flora, for example, before colonoscopy and female pelvic examination.

Another type of patient for whom transient bacteremia presents a special problem is the immunocompromised patient, whose immune host defense mechanisms do not function well. For example, they may have nonfunctioning phagocytic cells, may not produce antibodies, or may be receiving medical therapies that suppress immune function, such as steroids or cancer chemotherapy. Because they do not readily eliminate microorganisms that circulate during transient bacteremia, organisms that are harmless for the healthy host may emerge as significant pathogens to cause *opportunistic infections*.

Intermittent bacteremia, another important cause of bloodstream infection, occurs when organisms are intermittently released into the blood from a site of primary infection else-where in the body. The primary infections often result in acute, symptomatic disease character-ized by high fever, shaking chills, and cold and clammy skin. In severe cases, patients may become hypotensive and experience multiple organ-system failure, resulting in death. The most common sources of intermittent bacteremia are primary infections in the genitourinary tract (25%), respiratory tract (20%), abscesses (10%), surgical wounds (5%), biliary tract (5%), other known sites (10%), and uncertain sites (25%). The bacteria most commonly associated with in-termittent bacteremia are aerobic gram-negative bacilli, such as *Escherichia coli, Klebsiella pneumoniae*, and *Pseudomonas aeruginosa*, and various aerobic gram-positive cocci, especially *Staphylococcus aureus* and *Streptococcus pneumoniae*.

As the name implies, *continuous bacteremia* refers to constant or continual presence of bacteria in the bloodstream. Continuous bacteremia is a cardinal feature of endocarditis and other infections of the intravascular space. Implanted inanimate objects, such as intravascular catheters, shunts, and ports, may become infected, resulting in continual shedding of microorganisms into the bloodstream. Generally, patients with continuous bacteremia are symptomatic and require diagnosis of and treatment for their infection. The bacteria commonly associated with continuous bacteremia are viridans group streptococci (alpha-hemolytic) and *Staphylococcus epidermidis*.

Laboratory Diagnosis of Bacteremia

The laboratory diagnosis of bacteremia requires the recovery of a clinically significant microor-ganism in culture of the patient's blood. Thus, to establish the diagnosis, blood specimens must be properly collected and sent to the laboratory for appropriate culture. Health care professionals must be aware of their laboratory's principles and guidelines for optimum recovery of clinically significant microorganisms from a patient's blood. The three major guidelines are for: (1) skin antisepsis and venipuncture; (2) volume of blood collected for culture; and (3) number and tim-ing of blood cultures collected. These guidelines are reviewed briefly in the following sections.

Skin antisepsis and venipuncture.   Blood samples for culture are commonly collected by per-forming a venipuncture in the antecubital space of the arm. Because the skin has a normal micro-bial flora, a major problem with blood cultures is that they may be contaminated with skin flora, making the clinical significance of isolates difficult to interpret. This problem is best overcome by meticulous attention to skin antisepsis at the venipuncture site. Popular agents used for skin antisep-sis have included tincture of iodine or an iodophor, such as Betadine, but currently, tincture of chlorhexidine has replaced these agents at most health care institutions. This compound is easy to use and has improved effectiveness as a skin-sterilizing agent. It is **important** to note that before any blood sample is collected, the person collecting the sample, usually a phlebotomist, must properly identify the patient to verify that the specimen for culture is drawn from the correct patient.

In general, the venipuncture site should be cleansed with a 70% alcohol prep pad and then swabbed with the tincture of chlorhexidine pad by using concentric motions. After the chorhexidine has dried, the blood sample is collected by using a butterfly transfer set to collect blood directly into special blood culture bottles provided by the laboratory. In an alternative method, blood is collected into a Vacutainer tube containing an anticoagulant that is nontoxic to bacteria. The phlebotomist must ensure that all blood culture bottles or Vacutainer tubes used in the procedure are labeled with the patient's name. When the Vacutainer tubes are transported to the laboratory, the blood is transferred from the tubes into blood culture bottles. All blood culture bottles, whether inoculated directly or via a Vacutainer tube, are incubated at 35°C and observed for microbial growth.

Every effort must be made to adhere strictly to the guidelines for skin antisepsis established at each institution. Blood cultures that become contaminated because of the phlebotomist's failure to adhere to proper collection techniques may result in prolonged hospital stays, unnecessary administration of antibiotics (possibly leading to an increase in bacterial drug resistance), additional costs associated with performing repeat laboratory tests, and other associated costs. Several studies have shown that a contaminated blood culture may increase hospital costs by at least $10,000 for each patient. Because of this, hospitals and microbiology laboratories monitor blood culture contamination rates on a monthly basis. The nationally accepted average for a blood culture contamination rate is 3% or less. Because of the importance of this problem, any contamination rate higher than this value may result in remedial training and/or counseling of the individuals responsible for collecting the contaminated specimens.

Volume of blood cultured.   Patients with bacteremia may have very low numbers of bacteria circulating in their blood (sometimes less than 1 to 2 bacteria/ml). To detect bacteremia, at least one viable microorganism must be present in the specimen collected. In a particular hospital or health care institution, the microbiology laboratory provides recommendations as to the volume of blood collected for culture. Health care professionals must become familiar with and adhere to these guidelines. For adult patients, most laboratories recommend the collection of 10 to 20 ml of blood. This volume permits the inoculation of 5 to 10 ml into each of two blood culture bottles, one for cultivating aerobes (the aerobic bottle) and the other for anaerobes (the anaerobic bottle). Because of their smaller body size and because anaerobes are not important causes of bacteremia in pediatric patients, only a single aerobic bottle is inoculated with 1 to 5 ml of blood, depending on the size of the patient.

Timing and number of blood cultures.   The timing of blood specimens collected for culture is usually determined by whether the patient has intermittent or continuous bacteremia. If it is suspected that the patient has continuous bacteremia, then bacteria will always be present in the patient's blood, and the timing of specimen collection can be random. With intermittent bacteremia, however, the bacteria are present in the blood only occasionally, and therefore, the timing of the sample collection is of critical importance. If the patient is experiencing intermittent chills, blood should ideally be collected for culture one hour before the expected chill or temperature spike. There is usually a lag of about one hour between the influx of bacteria into the blood and the onset of chills. In practice, however, blood cultures are usually obtained after the onset of fever and chills, or they may be collected randomly over a 24-hour interval.

Guidelines also exist for the recommended number of blood specimens submitted for culture. Under no circumstances is it acceptable to collect only one blood specimen for culture. The reason is that it may be difficult to establish the clinical significance of the isolate if the organism recovered from culture is also a member of the normal skin flora. In some cases, skin flora bacteria, such as coagulase-negative staphylococci and *Propionobacterium acnes,* can cause significant human infections. If only one blood sample is collected, evaluating whether the isolate is clinically significant or represents a normal flora contaminant may be difficult. Thus, at least two, and usually not more than four, blood specimens should always be collected for culture. On occasion, more than four specimens may be needed to establish the laboratory diagnosis of infection, such as in those rare patients who may be experiencing very low levels of intermittent bacteremia or are infected with a microorganism that is especially difficult to isolate in culture.

Laboratory Testing and Reporting

Once the blood specimens for culture have been collected and received by the clinical microbiology laboratory, they are identified by patient name and date, and recorded, most often in a laboratory computer system. They are then placed in incubators at 35°C and examined periodically for the appearance of microbial growth. Most modern clinical microbiology laboratories use automated instruments with electronic sensors that monitor each blood culture bottle continually every 10 to 20 minutes to detect microbial growth. The BacTAlert instrument, distributed by bioMérieux in Durham, North Carolina, is one example of such an automated blood culture system. The instrument contains pull-out drawers in which inoculated blood culture bottles are placed in individual chambers (fig. 27.1a). Once the drawer is closed, the instrument electronically screens a sensor at the bottom of each bottle (fig. 27.1b), which changes color when carbon dioxide is produced during bacterial growth. When growth is detected, an alarm alerts the technologists that a blood culture in a specific location in the instrument is showing signs of bacterial growth. These instruments are so sensitive that they sometimes detect bacterial growth only a few hours after the blood culture bottle has been placed in the instrument. The medical technologist removes the indicated blood culture bottle from its specified location in the instrument, aseptically removes a broth sample from the bottle, and prepares a Gram-stained

Figure 27.1   (a) The BacTAlert automated blood culture instrument is illustrated here with the drawer containing blood culture bottles open at left. Once the drawer is closed, the instrument continuously monitors for the presence of microbial growth in the bottles and sends an alert when growth is detected. (b) Blood culture bottles for use with the system contain medium formulated to detect a variety of bacteria and yeast. The sensor at the bottom of the bottle changes color when carbon dioxide is released during bacterial growth, producing the positive growth signal.
(a) ©PJF Military Collection/Alamy Stock Photo; (b) ©Phanie/Alamy Stock Photo

(a)

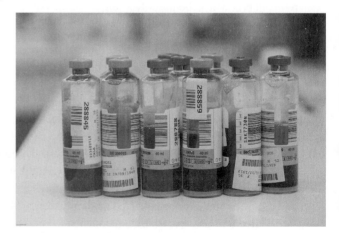

(b)

smear for examination. The Gram-stained smear result (e.g., gram-negative rods or gram-positive cocci in clusters resembling staphylococci) is then reported by telephone and by computer to the patient location. Because of the severity of bloodstream infections, in some laboratories the result is telephoned directly to the patient's physician. Most clinical microbiology laboratories are staffed 24 hours a day, 7 days a week, so that this continual monitoring and reporting system has an enormous beneficial impact on patient care. Physicians are alerted at the earliest possible time to the presumptive diagnosis of bacteremia and the need to administer appropriate antibiotics, when clinically indicated. Once the reporting activities are complete, the medical technologist proceeds to identify the organism and perform an antimicrobial susceptibility test, much as you have done in the preceding laboratory exercises. When all tests are completed, final reports of the microorganism's identity and antimicrobial susceptibility pattern are issued to the patient's doctor and location, but interim results are available when needed.

Molecular technologies, such as microarray assays, are being used increasingly in clinical microbiology laboratories for the rapid identification of bacteria directly in positive blood cultures. In addition, these assays can detect the presence of antibiotic-resistance genes, thus providing the physician with an early guide for antibiotic treatment. The results of the microarray assay can be available within one hour, and this has improved patient care. Microarray assays and their applications in the clinical microbiology laboratory are discussed in greater detail in Exercise 17.

Clinical Significance of Blood Culture Isolates (Interpretation of the Results)

In many cases, the genus and species identification of the microorganism, along with the patient's clinical presentation, indicate the significance of the isolate. For instance, the recovery of an organism such as *Neisseria meningitidis, Escherichia coli, Streptococcus pyogenes, Streptococcus pneumoniae*, or *Pseudomonas aeruginosa* from a blood culture establishes that the organism is involved in the patient's disease, because they are recognized pathogens associated with well-defined disease processes. The significance becomes even more clear if more than one blood culture is positive with the same organism. Sometimes, however, interpreting the clinical significance of a microorganism from a blood culture can be problematic, particularly when the organism is a member of the normal flora of the skin, such as coagulase-negative staphylococci. In these cases, the clinical significance of the isolate can usually be established if the same coagulase-negative staphylococcus is recovered from more than one blood culture. On the other hand, if several blood cultures are collected and only one culture grows a coagulase-negative staphylococcus, the isolate is generally regarded as a contaminant. This is why more than one blood culture should always be collected from a patient suspected of having bacteremia.

CASE STUDY

Three Patients with Positive Blood Cultures
Case 1.

A 50-year-old man complained of a low-grade fever (38°C) for 3 weeks with weight loss and loss of appetite, beginning about one week after dental extraction. As a child, he had a serious strep throat infection and was subsequently told he had a heart murmur. Three sets of blood culture were drawn, and all culture bottles became positive with gram-positive cocci in chains that were alpha-hemolytic on blood agar and **not** soluble in bile.

1. What is the most likely identification of the blood culture isolate?
2. What type of bacteremia did this person most likely have (transient, intermittent, continuous)?

3. Name the organism that caused the patient's strep throat initially. Does it have the same hemolytic reaction as the blood culture isolate?
4. What is the patient's clinical diagnosis?

Case 2.

A 25-year-old heroin abuser had abrupt onset of high fever (41°C), chills, and shortness of breath. Two of her three aerobic blood culture bottles became positive with a gram-negative bacillus that did not ferment glucose and was oxidase positive.

1. What is a possible genus identification of the organism?
2. What type of bacteremia did the patient most likely have?
3. Why did the organism grow only in the aerobic blood culture bottles?
4. What was the most likely source of this blood culture isolate?

Case 3.

A 60-year-old man convalescing in the hospital after bowel surgery still required intravenous fluids. He developed a low-grade fever (37.9°C) but had no specific complaints. One set of blood cultures was drawn and the aerobic bottle only grew gram-positive cocci in clusters that were negative with the latex agglutination test for *Staphylococcus aureus*.

1. What is the most likely identification of the organism grown from the blood culture bottle?
2. Does this patient have a bacteremia?
3. What is your impression of the significance of this blood culture organism? Does it require therapy?
4. What is the most likely source of the patient's isolate?

Questions

1. What is septicemia?

2. What is intermittent bacteremia, and how does it differ from transient bacteremia and continuous bacteremia?

3. Why should some patients with a history of rheumatic fever receive prophylactic antibiotic treatment before dental work?

4. What are the three most common primary sources of infection responsible for intermittent bacteremia?

5. Which agent is recommended for achieving excellent skin antisepsis before a blood sample is collected for culture? Why?

6. In the laboratory diagnosis of bacteremia, of what importance is the volume of patient blood collected?

7. In a patient suspected of having continuous bacteremia, what is the general recommendation for the timing of a blood culture collection? In a patient with intermittent bacteremia?

8. Is one blood culture considered acceptable for establishing the laboratory diagnosis of bacteremia? Why?

9. What is the importance of having the clinical microbiology laboratory telephone a preliminary report of the Gram-stained smear findings of a positive blood culture to the patient's physician and nursing floor?

10. What criteria are used in establishing the clinical significance of a blood culture isolate?

11. What is the value of using an automated blood culture system for detecting bacteremia?

PART FOUR

Microbial Pathogens Requiring Special Laboratory Techniques; Serodiagnosis of Infectious Disease; Principles and Practices of Infection Prevention

In Exercises 19 through 27 we have studied microbiological techniques for isolating and identifying aerobic or facultatively anaerobic bacteria. In this section we shall see how anaerobic bacteria are cultivated. The general techniques for identifying mycobacteria and microbial pathogens of other types (fungi, viruses, and animal parasites) are also described in these exercises. In some infections, evidence of the causative microorganism is obtained by examining the patient's serum for specific antibodies against the suspected pathogen. This is usually done when the microorganisms are difficult to cultivate by routine methods or sufficient time has elapsed such that the organism is no longer recoverable in culture from patient specimens. You should note that these techniques are quite varied. It is therefore important to remember that when specimens are ordered for laboratory diagnosis of microbial disease, the suspected clinical diagnosis should be stated on the request slip or in the computer so that appropriate laboratory procedures can be instituted.

In the final Exercise, we shall learn the principles and recommended practices for preventing transmission of infection in the health care environment.

Anaerobic Bacteria

Learning Objectives

After completing this exercise, students should be able to:
1. Discuss the basic principles of performing anaerobic methods in the laboratory.
2. Describe the types of clinical specimens that are acceptable for anaerobic culture.
3. Describe the types of clinical specimens that are unacceptable for anaerobic culture.
4. Specify the factors that predispose to the development of an anaerobic infection.
5. Name four clinically important species of clostridia and the diseases they cause.

Obligate (strict) anaerobic bacteria cannot grow in the presence of oxygen; therefore, in the laboratory, media containing reducing agents are used to cultivate anaerobes. Reducing agents such as sodium thioglycollate and cystine remove much of the free oxygen present in liquid media. Cooked meat broth is an excellent medium because it contains many reducing agents as well as nutrients. As for aerobic organisms, specialized enriched, selective, and differential agar media are available for cultivating anaerobic bacteria. Some of these are described in table 28.1. For the successful recovery of anaerobes from clinical specimens, conventional formulations of anaerobic media should be prepared, stored, and used under oxygen-free conditions to prevent the formation of oxidized products that may be toxic or inhibitory to the growth of anaerobes in a clinical specimen. However, new formulations of anaerobic media are now available that do not require oxygen-free preparation and storage conditions.

Table 28.1 Examples of Culture Media for the Isolation and/or Presumptive Identification of Pathogenic Anaerobic Bacteria from Clinical Specimens

Media	Classification	Enrichment, Selective and/or Differential Agent(s)	Type(s) of Organisms Isolated
Thiogycollate broth	All purpose	Cystine, yeast extract, sodium thioglycollate	Anaerobes and aerobes
Cooked meat medium	Enriched, differential	Heart tissue granules Peptic digest of animal tissue	Clostridium spp. (detects proteolysis) Non-endospore-forming anaerobes
CDC anaerobe blood agar medium	Enriched	Yeast extract, vitamin K_1, hemin, cysteine	Most anaerobes, especially fastidious and slow growing
Phenylethyl alcohol agar (PEA)	Selective	Phenylethyl alcohol (inhibits gram-negatives)	Anaerobic cocci
KV medium	Enriched, differential Selective	Laked blood (frozen and thawed, lysed red blood cells), vitamin K_1, hemin Vancomycin (inhibits gram-positive bacteria), kanamycin (inhibits facultative gram-negative bacilli)	Peptococcus melaninogenicus (produces black pigment on this medium)
Egg Yolk agar (also called lecithinase-lipase agar)	Enriched, Differential	Vitamin K_1, hemin Egg yolk*	Presumptive identification of Clostridium perfringens

*Egg yolk contains lecithin and lipase. The degradation of lecithin in the egg yolk by the organism results in an opaque precipitate around the colonies (see fig. 28.3, left image). Lipase enzyme hydrolyzes the fats within the egg yolk, resulting in an iridescent sheen on the colony surface. Proteolysis of the egg yolk may also be observed as indicated by a clearing of the medium around the colonies.

To ensure complete removal of oxygen from the culture environment, cultures are incubated in an "anaerobic jar," of which there are two types. One type has a lid fitted with an outlet through which air can be evacuated by a vacuum pump and replaced by an oxygen-free gas. A catalyst in the lid catalyzes the reduction and chemical removal of any traces of oxygen that may remain. The other type also requires use of a catalyst in the jar. A foil envelope containing substances that generate hydrogen and CO_2 is placed in the jar with the cultures. The envelope is opened, and 10 ml of tap water are pipetted into it. When the jar is closed (the lid is clamped down tightly), the hydrogen given off combines with oxygen, through the mediation of the catalyst, to form water. The CO_2 helps to support growth of fastidious anaerobes. A second envelope placed in the jar with the first contains a pad soaked with an oxidation-reduction indicator, for example, methylene blue. When the pad is exposed, the color of the dye indicates whether or not oxygen is present in the jar atmosphere; methylene blue is colorless in the absence of oxygen, blue in its presence. Figure 28.1 illustrates one brand of anaerobic jar (GasPak, B-D Diagnostic

Figure 28.1 A GasPak jar (BD Diagnostic Systems) for cultures to be incubated anaerobically. In this model, the tight-fitting lid contains a catalyst. The large foil envelope has been opened to receive 10 ml of water delivered by a pipette. With the lid clamped in place, hydrogen generated from substances in the large envelope combines with oxygen in the jar's atmosphere. This combination is mediated by the catalyst and forms water, which condenses on the sides of the jar. Carbon dioxide is also given off by the substances within the large envelope, contributing to the support of growth of fastidious organisms. The smaller envelope has also been opened to expose a pad (arrow) soaked in methylene blue, an indicator used to detect the presence or absence of oxygen. When first exposed, the pad was blue in color. Now it is colorless, indicating that there is no free oxygen left in the jar, for the indicator dye loses its color in the absence of oxygen. The jar now contains an anaerobic atmosphere and can be placed in the incubator. ©*Josephine A. Morello*

Systems) in use. Yet another type of anaerobic system consists of a plastic pouch into which up to four Petri plates can be placed (fig. 28.2). An anaerobic gas-generating sachet is placed in the bag with the plates. The sachet absorbs oxygen from the pouch and generates CO_2 without the need for water. This type of system is convenient when only small numbers of plates are to be incubated anaerobically. In addition, the plates can be viewed for growth through the transparent plastic pouch without exposing the organisms to oxygen. Once sufficient growth is observed, the plates can be opened and the organisms identified.

In recent years, improved techniques for anaerobic culture have been developed in response to an increasing interest in the role of anaerobic bacteria as agents of human infections. Anaerobes have been implicated in a wide variety of infections (see table 28.2) from which multiple species of bacteria are recovered in culture—that is, mixed infections.

The ability of anaerobic organisms to grow in and damage body tissues depends on how well the tissues are oxygenated. Any condition that reduces their oxygen supply, making them *anoxic* (without oxygen), provides an excellent environment for the growth of anaerobes. Impairment of local circulation because of a crushing wound, hematoma, or other compression leads to tissue anoxia and sets the stage for contaminating anaerobes, if they are present, to cause human infection. Except for those infections caused by members of the genus *Clostridium,* most anaerobic infections are characterized by the presence of a foul smell. The odor results from growth of anaerobes in the anoxic tissues where they produce gases, short-chain fatty acids, and alcohols as end products of metabolism. These compounds are ones we associate with the process of rotting or putrefaction.

Numerous genera of anaerobic bacteria have been recognized as pathogens, or potential pathogens, and almost all of them are members of the body's normal flora. The significance of their isolation from a clinical specimen may be difficult to determine, and their pathogenicity is not yet fully understood. Generally, they appear to be "pathogens of opportunity"—that is, given the opportunity to gain access to tissue with impaired blood supply, they may grow and cause enough tissue destruction to establish a local infection. The extent of local or systemic damage may be related to a number of factors, including the properties of microorganism(s) involved, the initial site of infection, and the defense mechanisms of the infected individual. Because anaerobes are part of the normal flora of the body, physicians may have difficulty assessing their importance in a culture taken from an infected area. They must consider whether or not a given isolate may be merely a contaminant from the local normal flora as they evaluate a patient's clinical condition. The microbiologist can assist by offering directions for the proper collection and transport of specimens,

Figure 28.2 The anaerobe pouch is a convenient system to use when only small numbers of plates are to be incubated anaerobically. The plates are put into the pouch with a gas-generating sachet and an oxygen-reduction indicator. ©*Josephine A. Morello*

Microbial Pathogens Requiring Special Laboratory Techniques

Table 28.2   Infections in Which Anaerobic Bacteria Have Been Implicated

Body Area Involved	Type of Infection
Infections of the female genital tract	Endomyometritis
	Salpingitis
	Peritonitis
	Pelvic abscess
	Vaginal abscess
	Vaginitis
	Bartholin's abscess
	Surgical wound infection
Intra-abdominal infections	Peritonitis
	Visceral abscess
	Intraperitoneal abscess
	Retroperitoneal abscess
	Traumatic or surgical wound infection
Pleuropulmonary infections	Pneumonia
	Lung abscess
	Empyema
Miscellaneous infections	Osteomyelitis
	Cellulitis
	Arthritis
	Abscesses
	Brain abscess
	Endocarditis
	Meningitis

reporting on the predominance of microorganisms present in a culture, and assuring adequate identification of any significant anaerobes that may be isolated.

Most anaerobic infections can be treated with penicillin, because the majority of anaerobes have not yet developed resistance to this antibiotic. A notable exception is a species of *Bacteroides, Bacteroides fragilis,* which causes intra-abdominal and gynecologic infections and is typically resistant to penicillin. Clindamycin is another drug commonly used for treating many anaerobic infections, including those caused by *B. fragilis.* Unfortunately, with use of this drug, there is a high risk that the patient may develop antibiotic-associated diarrhea, as will be discussed later.

Some of the important genera of anaerobic bacteria are listed in table 28.3, most of which are either gram-positive or gram-negative, nonsporing bacilli. However, certain *Clostridium* species, which are gram-positive *endospore-forming* bacilli, are recognized pathogens and are best understood. Many species in the genus *Clostridium* are commonly found in the intestinal tract of humans and animals, as well as in the soil, but four are of particular importance in human disease: *C. perfringens, C. tetani, C. botulinum,* and *C. difficile.* Each of these is associated with a different and characteristic type of clinical disease.

Clostridium perfringens is an agent of *gas gangrene* (sometimes in association with other clostridia). If they gain entry into a deep tissue wound with impaired blood flow, they can multiply rapidly, using tissue carbohydrates, liberating gas, and producing enzymes that cause additional destruction that makes more nutrient available to the bacteria. This can quickly develop into a life-threatening situation if not promptly treated by surgical debridement (removal of dead tissue) and aeration of the injured tissue, and by antimicrobial agents. A *Clostridium perfringens*

Table 28.3 Some Important Genera of Anaerobic Bacteria

Basic Morphology	Genera	Infections
Bacilli		
Gram-positive, endospore-forming	Clostridium	C. perfringens—gas gangrene C. tetani—tetanus C. botulinum—botulism C. difficile—antibiotic-associated diarrhea
Gram-positive, nonsporing	Actinomyces	Actinomycosis
	Eubacterium	Infections of female genital tract, intra-abdominal infections, endocarditis
	Propionibacterium	Difficult to assess; has had clinical significance in cultures of blood, bone marrow, and spinal fluid
	Bifidobacterium	Occasionally isolated from blood; significance not established
Gram-negative, nonsporing	Bacteroides	Infections of female genital tract, intra-abdominal and pleuropulmonary infections; well-established as a pathogen
	Fusobacterium and Prevotella	Same as Bacteroides but less frequent
	Leptotrichia	Found in mixed infections in oral cavity or urogenital areas; significance not established
Cocci		
Gram-positive	Peptostreptococcus (anaerobic streptococci)	Infections of female genital tract, intra-abdominal and pleuropulmonary infections; often found with Bacteroides; established pathogen
Gram-negative	Veillonella	Found in mixed anaerobic oral and pleuropulmonary infections; significance not established

enterotoxin is also a common cause of food-borne illness, resulting primarily from ingestion of improperly cooked and stored beef. *C. perfringens* grows well on blood agar plates in an anaerobic jar, showing characteristic double zones of hemolysis, and is positive for lecithinase but negative for lipase on egg yolk agar medium (see fig. 28.3). Its enzymes attack the proteins and carbohydrates of milk, producing "stormy fermentation" of a milk medium, with clotting and gas formation. It ferments a number of carbohydrates with the production of acid and gas. Usually it does not form endospores in ordinary culture media, nor does it do so when growing in tissues.

Clostridium tetani is the agent of *tetanus,* or "lockjaw." When introduced into deep tissues, this organism produces little or no local tissue damage, but it secretes an exotoxin known as a neurotoxin, which adversely affects nerve function. The neurotoxin is absorbed from the infected area and travels along peripheral motor nerves to the spinal cord. Severe muscle spasm and convulsive contraction of the involved muscles result. It is often difficult to make a laboratory diagnosis of this disease because the site of injury may be closed and healed, with no apparent signs of infection, by the time the symptoms of neurotoxicity begin. The organism is difficult to cultivate, but if isolated, it is identified by microscopic morphology and patterns of carbohydrate fermentation. The endospore of *C. tetani* is usually at one end of the bacillus (terminal). It is wider in diameter than the vegetative bacillus, giving the cell the appearance of a "drumstick" or tennis racquet. The diagnosis is usually based on clinical signs and symptoms characteristic of the disease.

Clostridium botulinum produces an exotoxin that causes the deadly form of food poisoning called *botulism.* This is not an infectious disease, but a toxic disease. If the endospores of this soil organism survive in processed foods that have been canned or vacuum packed, they may multiply in the anaerobic conditions of the container, elaborating their potent exotoxin in the process of growth. If the food is eaten without further cooking (which would destroy the toxin), botulinum toxin enters the bloodstream from the gastrointestinal tract and travels to the

Microbial Pathogens Requiring Special Laboratory Techniques

Figure 28.3 These culture effects demonstrate some of the enzymes produced *by Clostridium perfringens,* which are important in its ability to destroy tissue. On the left (a) the activity of the alpha toxin, also known as lecithinase, is shown when *Clostridium perfringens* grows on egg yolk agar. The opaque precipitate results from the lecithinase activity. On the right (b), the typical double zone of hemolysis produced on a blood agar plate is seen. The theta toxin is responsible for the narrow zone of complete hemolysis adjacent to the colony, and the alpha toxin (lecithinase) is responsible for the wider zone of incomplete hemolysis.
(a) ©Josephine A. Morello; (b) ©John Prescott, University of Guelph

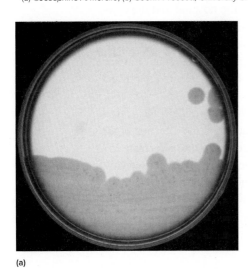

(a)

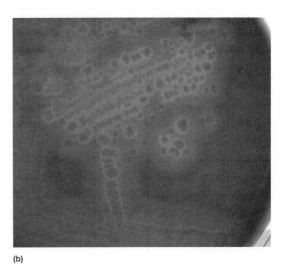

(b)

neuromuscular junctions of skeletal muscle—the site of action. There it prevents the release of a neurotransmitter, acetylcholine, whose effect is to initiate the signal for muscle contraction. In the absence of muscle contraction, the symptoms of botulism include double vision and difficulty swallowing, and if the disease is unrecognized and untreated, it can proceed to muscular paralysis and respiratory failure. The effects of this toxin on nerves and muscles are responsible for its medicinal use as Botox, a highly dilute preparation of botulinum toxin A, in people with facial spasms, palsy, wrinkles, and excessive sweating, for example.

Botulism is difficult to diagnose bacteriologically in culture. However, specialized laboratory tests can be used to demonstrate the presence of toxin in the incriminated food or the patient's blood or stool.

In *infant botulism,* when endospores of the bacillus (endospores in honey have been implicated in a few cases) are ingested by children under one year of age, the endospores germinate in the child's intestinal tract in some instances, and the resulting vegetative cells produce toxin. This type of botulism has been implicated in rare cases of sudden infant death syndrome (SIDS).

Clostridium difficile causes *antibiotic-associated diarrhea,* a condition that can progress to a more severe, life-threatening disease called *pseudomembranous colitis.* The organism causes disease by producing an enterotoxin and a cytotoxin that, respectively, cause intracellular fluid loss and cellular damage to the intestinal mucosa. In some individuals, *C. difficile* is found in low numbers as normal flora of the bowel, and its endospores are common environmental contaminants of patient rooms in hospitals and nursing homes. The use of certain antibiotics to which *C. difficile* is resistant, such as clindamycin, can disrupt the normal flora of the bowel, resulting in overgrowth of the resistant strains. Toxin production is then responsible for the disease (i.e., antibiotic-associated diarrhea). Alternatively, following antibiotic use and ingestion of endospores in the hospital environment, the disrupted bowel flora permit colonization and subsequent overgrowth of *C. difficile,* resulting in disease. *C. difficile* is recognized as the leading cause of *hospital-associated,* or *nosocomial,* diarrhea.

Anaerobic Bacteria

C. difficile also has emerged as an important cause of community-acquired, antibiotic-associated diarrhea in people who have never been hospitalized but have received a course of antibiotic therapy that predisposes them to infection. The problem became further complicated in 2005 by the emergence of a mutant strain of *C. difficile* that produces 25 to 30 times more than the normal amounts of enterotoxin and cytotoxin. This new strain causes more serious disease in humans, resulting in a higher mortality rate. It is often referred to as hypervirulent *C. difficile* or the NAP1 strain.

Given the widespread incidence of this serious disease in the hospital and the community, laboratory methods are needed for the rapid and reliable diagnosis of infection. These methods would ensure that appropriate therapy is instituted for the management of the infection. Reliable laboratory diagnostic methods are particularly needed for hospitalized and nursing home patients so that appropriate infection prevention measures can be implemented to prevent the transmission of infection to other patients.

Several types of tests have been developed for the laboratory diagnosis of *C. difficile* infection, ranging from culture of the organism from the stool of infected patients to the detection of the enterotoxin and/or cytotoxin in stool by cell culture assay or enzyme immunoassays (see **colorplate 25**). Each of these tests has inherent advantages and disadvantages with no test being ideal. Recently, however, gene amplification assays, such as PCR, have been developed, which provide a highly sensitive and specific method for the reliable diagnosis of infection. Evaluation of these gene amplification assays shows that they approach nearly 100% in their ability to detect infection.

In this exercise, we shall use species of *Clostridium* to illustrate the general principles of anaerobic culture methods.

Purpose	To learn basic principles of anaerobic bacteriology
Materials	Anaerobe jar
	Blood agar plates
	Thioglycollate broth
	Tubed skim milk
	Phenol red broths (glucose, lactose)
	Blood agar plate cultures of *Clostridium perfringens* and *Clostridium histolyticum*
	Nutrient slant cultures of *Pseudomonas aeruginosa* and *Staphylococcus epidermidis*
	Wire inoculating loop
	Marking pen or pencil
	Test tube rack

Procedures

1. Make a Gram stain of each *Clostridium* culture.
2. Select one of the *Clostridium* cultures and inoculate it on each of two blood agar plates. Label one plate "aerobic," the other "anaerobic."
3. Inoculate a tube of thioglycollate broth with *C. perfringens*. Inoculate a second tube of this medium with *P. aeruginosa*, and a third with *S. epidermidis*.
4. Inoculate a tube of milk with *C. perfringens* and a second milk tube with *C. histolyticum*.
5. Inoculate each *Clostridium* culture into phenol red glucose and lactose, respectively.
6. Incubate the blood agar plate labeled "aerobic" in air at 35°C for 24 hours.
7. Place all other tubes and plates in the anaerobe jar. (The instructor will demonstrate the method for obtaining an anaerobic atmosphere within the jar.) When it has been set up, the jar is incubated at 35°C for 24 hours. When working with actual clinical specimens, the jar is often not opened until 48 hours, except when *Clostridium* is highly suspected.

Results

1. Indicate the Gram reaction of the *Clostridium* cultures and illustrate their microscopic morphology.

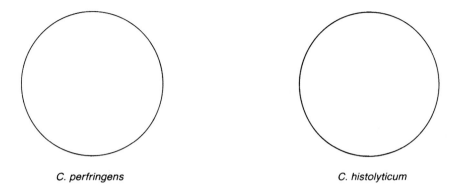

C. perfringens   C. histolyticum

What stain would you use to determine whether these organisms had produced endospores?

_____

Would you expect to find endospores in the blood agar plate culture of a *Clostridium*? _____

Why? _____

2. Examine the thioglycollate broth cultures (do not shake them). On the following figures make a diagram of the distribution of growth in each tube.

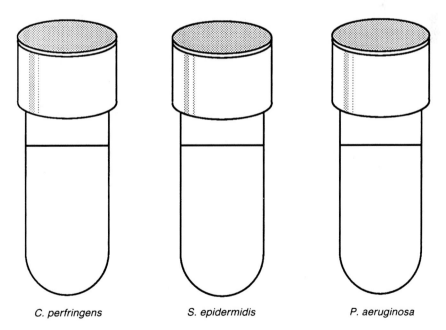

C. perfringens   S. epidermidis   P. aeruginosa

What is your interpretation of the appearance of these tubes?

3. Examine the blood agar plate cultures, milk tubes, and phenol red carbohydrate broths. Record your observations.

Name of Organism	Morphology on Aerobic Plate	Morphology on Anaerobic Plate	Hemolysis	Milk	Glucose	Lactose
C. perfringens						
C. histolyticum						

State your interpretation of the appearance of the milk cultures.

CASE STUDY

A Case of Diarrhea in a Hospitalized Patient

A 63-year-old hospitalized woman underwent a hysterectomy for cervical cancer and was treated for 10 days with the antibiotics clindamycin and ceftriaxone. Two days after the antibiotic treatment was discontinued, she developed profuse, watery diarrhea; cramping abdominal pain; and fever. A laboratory test revealed the presence of bacterial enterotoxin and cytotoxin in her stool sample.

1. What bacterium is causing this patient's infection?
2. What is the name of her disease?
3. How did she acquire this infection?
4. Which antibiotic is most likely responsible for this patient's infection? Why?
5. Can this infection be transmitted to other people in her room? By what means?

Microbial Pathogens Requiring Special Laboratory Techniques

Questions

1. Define anaerobe, aerobe, and facultative anaerobe.

2. Name examples of nonselective, selective, and differential media for isolating anaerobes.

3. Describe two methods for obtaining an anaerobic atmosphere for cultures.

4. Can a strict aerobe be distinguished from an anaerobe in thioglycollate broth? If so, how?

5. If you wanted to culture a wound specimen but couldn't find an anaerobe jar, would a candle jar serve as a suitable substitute? Why?

6. What is a bacterial endospore? Why should it have medical importance?

7. What is an exotoxin? What is a neurotoxin?

8. Describe two important properties of *C. perfringens* in culture.

9. Name three diseases caused by anaerobic bacteria.

10. When a specimen from a wound of a patient suspected of having gas gangrene is sent to the laboratory, would an immediate Gram-stain report be of clinical value? Why?

11. If a patient on a surgical unit develops gas gangrene, what hospital precautions, if any, should be taken? Why?

12. What is an opportunistic pathogen (pathogen of opportunity)?

13. Is botulism considered to be an infectious disease? Why?

14. Why should a cook follow home-canning instructions carefully?

15. What organism causes antibiotic-associated diarrhea? How does it produce disease?

16. What laboratory test is most reliable for diagnosing infection caused by the organism in question 15?

Mycobacteria

Learning Objectives

After completing this exercise, students should be able to:

1. Name a medium that is used to cultivate mycobacteria.
2. Explain the importance of digesting and concentrating a sputum sample before performing culture for mycobacteria.
3. Discuss the advantages of using the fluorescent stain for visualizing mycobacteria in clinical specimens.
4. Explain the basis of the PPD skin test.
5. Describe the type of patient for whom the QuantiFERON TB Gold In-Tube Test is used.

The genus *Mycobacterium* contains many species, a number of which can cause human disease. A few are saprophytic organisms, found in soil and water, and also on human skin and mucous membranes. The two most important pathogens in this group are *Mycobacterium tuberculosis,* the agent of tuberculosis, and *Mycobacterium leprae,* the cause of Hansen disease (leprosy). However, *Mycobacterium kansasii* and the *Mycobacterium avium* complex (see table 29.1) cause disease in persons with chronic lung disease and are being seen more frequently as opportunistic pathogens in patients with leukemia and acquired immunodeficiency syndrome (AIDS). Table 29.1 summarizes some of the mycobacteria that are human pathogens, according to the type of disease that they may cause. Species of mycobacteria that are commensals and not normally associated with human disease are listed as well.

Table 29.1   Mycobacteria in Infectious Disease

Disease	Species	Host(s)	Route of Entry
Tuberculosis	*Mycobacterium tuberculosis** *M. bovis**	Human Cattle and human	Respiratory Alimentary (milk)
Pulmonary disease resembling tuberculosis (mycobacterioses)	*M. avium** *M. intracellulare** *M. kansasii* *M. szulgai* *M. xenopi*	Fowl and human Human Human Human Human	Respiratory Environmental contacts? (water and soil)
Lymphadenitis (usually cervical)	*M. tuberculosis* *M. scrofulaceum* *M. avium* complex	Human Human Fowl Human	Respiratory Environmental contacts? (water and soil)
Skin ulcerations	*M. ulcerans* *M. marinum*	Human Fish and human	Environmental contacts? Aquatic contacts
Soft tissue	*M. fortuitum* *M. chelonae*	Human Human	Environmental contacts?
Hansen disease (leprosy)	*M. leprae*	Human	Respiratory
Saprophytes: water, soil; human skin and mucosae (normally do not produce disease)	*M. smegmatis* *M. phlei* *M. gordonae*		

*Mycobacterium tuberculosis and *M. bovis* are so closely related that they are often identified as the *Mycobacterium tuberculosis* complex. In like manner, *M. avium* and *M. intracellulare* are referred to as the *M. avium* complex.

The conventional laboratory diagnosis of tuberculosis and other mycobacterial infection is made by identifying the organisms in acid-fast smears and in cultures of clinical specimens from any area of the body where infection may be localized. More recently, molecular-based methods, such as nucleic acid probes and polymerase chain reaction (PCR) (see Exercise 17), have been developed. These methods characterize mycobacteria to the species level once they are recovered in culture, or identify them directly in the clinical sample without the need to perform culture. The availability of these methods has greatly reduced the time required for establishing the laboratory diagnosis of mycobacterial infection, which allows for early administration of appropriate therapy. In pulmonary disease, sputum specimens and gastric washings are appropriate, but if the disease is disseminated, the organisms may be found in a variety of areas. Urine, blood, spinal fluid, lymph nodes, or bone marrow may be of diagnostic value, especially in immunocompromised patients. Any specimen collected for identification of mycobacteria must be handled with particular caution and strict asepsis. These organisms, with their thick waxy coats, can survive for long periods even under adverse environmental conditions. They can remain viable for long periods in dried sputum or other infectious discharges, and they are also resistant to many disinfectants. An effective disinfectant for chemical destruction of tubercle bacilli must be chosen carefully.

Clinical samples such as sputum must be digested and concentrated before they are cultured for mycobacteria. During the digestion process, the thick, viscous sputum is liquefied so that any mycobacteria present, particularly if present in low number, are distributed evenly throughout the specimen. Digestion also kills other bacteria that would quickly overgrow in culture any mycobacteria present. After digestion, the sample is centrifuged at high speed to concentrate the mycobacteria at the bottom of the centrifuge tube. The supernatant fluid is discarded into an appropriate disinfectant, and an acid-fast-stained smear of a portion of the centrifuged pellet is prepared and examined microscopically. Another portion is cultured on special mycobacterial culture media. The digestion and concentration steps greatly improve the microscopic detection of mycobacteria in clinical specimens and their recovery in culture.

The Kinyoun stain (see Exercise 5) is commonly used to stain acid-fast bacilli. The organisms stain red against a blue background, whereas non-acid-fast organisms are blue (see **colorplate 9**). Tubercle bacilli are slender rods, often beaded in appearance. Another stain uses the fluorescent dyes auramine and rhodamine, which permit rapid detection of the organisms as bright rod-shaped objects against a dark background (see **colorplate 9**). Smears stained with fluorescent dyes are easier to read than are those stained by the carbolfuchsin method. Fluorescing organisms stand out prominently against the black background, and the smears can be examined microscopically at a lower magnification than smears stained with carbolfuchsin. In this way, a larger area of the smear can be examined in a shorter time period. This technique is used in many laboratories today.

Most mycobacteria do not grow on conventional laboratory media, such as chocolate or blood agar plates. Therefore, special solid media containing complex nutrients, such as eggs, potato, and serum, are used for culture. Lowenstein-Jensen medium is a solid egg medium prepared as a slant and is one of several in common use (see **colorplate 40**). Broth media are also available.

Tubercle bacilli and most other mycobacteria grow very slowly. At least several days, and up to 4 to 8 weeks for *M. tuberculosis,* are required for visible growth to appear. They are aerobic organisms, but their growth can be accelerated to some extent with increased atmospheric CO_2. Automated instruments that detect the CO_2 released by mycobacteria growing in liquid media are used in many laboratories and permit detection before visible growth is apparent. The CO_2 is released when mycobacteria metabolize special substrates present in broth culture media.

Tuberculosis is an important infectious disease, with more than one-third of the world's six billion population infected with *M. tuberculosis.* Fortunately, more than 90% of the infected people are asymptomatic and do not serve as an infectious risk for disease transmission. However, infected people carry the organism in a "dormant" state within their bodies, and the disease can reactivate to cause symptomatic tuberculosis. Reactivation usually occurs in those infected persons who become immunocompromised by certain conditions. Upon reactivation, the disease can be fatal and can be transmitted to other individuals who have close contact with the symptomatic person.

Skin testing with an antigen derived from *M. tuberculosis*, called purified protein derivative or PPD, is an effective way to identify individuals who have symptomatic or asymptomatic disease. *M. tuberculosis* infection results in the sensitization of a certain type of lymphocytes in the body, which are associated with a delayed-type hypersensitivity reaction (also known as Type IV hypersensitivity). If the patient has been infected with *M. tuberculosis*, when the test is performed by the injection of PPD into the skin with a tuberculin syringe, a zone of induration appears after 48 to 72 hours (see **colorplate 41**). The zone of induration is caused by the influx of lymphocytes and macrophages to the area where the PPD was injected. If the zone size is greater than a certain diameter, it is considered a positive test for tuberculosis, indicating that the person has been infected. If a previous skin test was negative and a subsequent skin test is positive, the person is classified as a converter. In this case, he or she would receive antituberculosis therapy to kill the dormant mycobacteria present in the body and prevent reactivation of tuberculosis.

The PPD skin test is useful for screening individuals who have normal immune function and have not been vaccinated against tuberculosis. Reliable skin testing cannot be performed on persons with abnormal immune function (e.g., human immunodeficiency virus [HIV]-infected patients, because of possible defects in lymphocyte function) or people who have received the tuberculosis vaccine. The vaccine, BCG (Bacille Calmette Guerin), consists of a live, weakened strain of *Mycobacterium* that is closely related to *M. tuberculosis*. Persons who receive the BCG vaccine become sensitized to PPD. Thus, BCG-vaccinated people should not receive a PPD skin test, because they will have a "false"-positive skin test reaction and may even experience a more serious delayed hypersensitivity reaction. In BCG-vaccinated individuals, the QuantiFERON TB Gold In-Tube Test can be used, because it tests for the patient's immunity to antigens that are specific for *M. tuberculosis*. This test is not a skin test but uses whole blood to detect cellular immunity. Although it does not give false-positive reactions, it has not replaced the PPD skin test for testing non-BCG-vaccinated patients, because of its expense.

M. leprae (also known as Hansen bacillus) cannot be cultivated on laboratory media. Laboratory diagnosis of Hansen disease is based only on direct microscopic examination of acid-fast smears of material from the lesions.

EXPERIMENT **29.1**   **Microscopic Morphology of *Mycobacterium tuberculosis***

Purpose	To study *M. tuberculosis* in smears
Materials	Prepared acid-fast stains

Procedures

Examine the prepared slides under oil immersion. Make a colored drawing of tubercle bacilli as you see them.

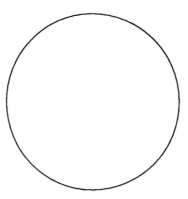

State your interpretation of the term "acid-fast."

EXPERIMENT **29.2**   **Culturing a Sputum Specimen for Mycobacteria**

Purpose	To study mycobacteria in culture
Materials	Lowenstein-Jensen slants
	Simulated sputum culture (predigested and concentrated)
	Marking pen or pencil

Procedures

1. Prepare an acid-fast stain directly from the sputum specimen (review Exercise 5). Read and record observations.
2. Inoculate the specimen on Lowenstein-Jensen medium, label the slant, and incubate at 35°C until growth appears.
3. Examine for visible growth and record appearance.

Results

1. Diagram observations of the stained smear.

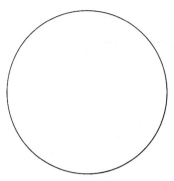

2. Describe the growth on Lowenstein-Jensen medium. How many days before growth appeared?

CASE STUDY

A Case of Pneumonia in a Patient with Chronic Alcoholism

A 72-year-old male with a long history of alcoholism went to the emergency room for evaluation of a progressively worsening cough that was producing blood-tinged sputum (hemoptysis). The patient complained of frequent night sweats and a 25-pound weight loss over the past 2 months. His medical record indicated that he had received a PPD skin test one year previously, which was interpreted as "positive," but the patient refused treatment at that time. A sputum sample now sent to the microbiology laboratory revealed the presence of acid-fast bacilli when a Kinyoun stain was performed.

1. What disease does this man have and which microorganism is most likely responsible for his infection?
2. Did the patient's long history of alcohol abuse contribute to the development of his disease?
3. How did the patient most likely acquire his infection?
4. What culture media must be used to grow this organism in the laboratory?
5. What was the significance of the patient's positive PPD skin test?

Questions

1. Name two saprophytic, commensal species of *Mycobacterium.*

2. What is Hansen bacillus? Name the genus and species of the responsible organism.

3. Explain why sputum is "digested" and concentrated before culture.

4. Why are tubercle bacilli acid-fast?

5. Why are tubercle bacilli difficult to destroy by chemical disinfection?

6. What special precautions are necessary for collecting and handling specimens from tuberculosis patients?

7. Is the presence of acid-fast bacilli in a sputum sample sufficient evidence of tuberculosis? Why?

8. Why do some culture reports for pathogenic mycobacteria require 6 or more weeks?

9. If you were caring for a patient with tuberculosis, what type of isolation precautions would you use?

10. In the United States, an acid-fast isolate from a patient with AIDS is most likely to be which mycobacterial species?

11. Name two screening tests that can be used to determine whether a patient is infected with *M. tuberculosis*.

Mycoplasmas, Rickettsiae, Chlamydiae, Viruses, and Prions

Learning Objectives

After completing this exercise, students should be able to:
1. Distinguish between mycoplasmas and other bacteria.
2. Explain how rickettsia and ehrlichia infections are transmitted to humans.
3. Discuss the differences between a virus and a bacterium.
4. Describe various methods for the laboratory diagnosis of viral infections.
5. Define prions and specify the diseases they cause.

Mycoplasmas, rickettsiae, and chlamydiae are classified as true bacteria, but they are extremely small, and for various reasons cannot be cultured by ordinary bacteriologic methods. The viruses are the smallest of all microorganisms and are classified separately. The techniques that have developed over many years for propagating and studying viruses have provided an understanding of their nature and pathogenicity. The electron microscope, together with elegantly precise biochemical, physical, molecular, and immunologic procedures, has revealed the structure of viruses and their role in disease at the cellular level. Prions are proteinaceous infectious particles that cause so-called slow viral infections because they take many years to develop. Prions are smaller than viruses and are believed to contain no nucleic acids. The means by which such agents can cause disease is under intensive investigation, and ongoing molecular studies may unravel the answer.

In this exercise we shall review the nature and pathogenicity of these microorganisms.

Purpose	To learn the role of mycoplasmas, rickettsiae, chlamydiae, viruses, and prions in disease and to review some laboratory procedures for recognizing them
Materials	Audiovisual or reading materials illustrating each group Diagram of the electron microscope

Procedures

Students will not perform laboratory procedures, but they should come to class prepared by assigned reading to discuss the laboratory diagnosis of diseases caused by these agents.

Following is a brief summary of each group.

Mycoplasmas

The mycoplasmas, previously called "pleuropneumonia-like organisms" (PPLO), were first known as etiologic agents of bovine pleuropneumonia. Several species are now recognized, including three that are agents of human infectious disease.

Mycoplasma pneumoniae is the causative organism of "primary atypical pneumonia." The term implies that the disease is unlike bacterial pneumonias and does not represent a secondary infection by an opportunistic invader, but has a single primary agent. Clinically, mycoplasmal pneumonia resembles an influenza-like illness.

Mycoplasma hominis may be found on healthy mucous membranes, but it is also associated with some cases of postpartum fever, pyelonephritis, wound infection, and arthritis.

Ureaplasmas are strains of mycoplasma that produce very tiny colonies and were, for this reason, once called "T-mycoplasmas." They have been renamed in recognition of their unique possession of the enzyme urease. These mycoplasmas, like *M. hominis,* are normally found on mucosal surfaces, but they have sometimes been associated with urogenital and neonatal infections and female infertility.

Mycoplasmas are extremely pleomorphic (varied in size and shape). They are very thin and plastic because they lack cell walls. For this reason, unlike other bacteria, they can pass through bacterial filters, they do not stain with ordinary dyes, and they are resistant to antimicrobial agents (such as penicillin) that act by interfering with cell wall synthesis.

Laboratory Diagnosis

These organisms can be cultivated on enriched culture media, but on agar media their colonies can be clearly seen only with magnifying lenses. They do not heap on the surface but extend into and through the agar from the point of inoculation.

Specimens for laboratory diagnosis by culture include sputum, urethral or cervical discharge, synovial fluid, or any material from the site of suspected infection. Cultures require 3 to 10 days of incubation at 35°C. Serological methods can be used to detect mycoplasmal antibodies in the patient's serum, and nucleic acid amplification tests, such as PCR, have become available for the rapid and reliable diagnosis of *M. pneumoniae* respiratory infections.

Rickettsiae

The rickettsiae are very small bacteria that survive only when growing and multiplying intracellularly in living cells. In this respect they are like viruses, that is, they are obligate intracellular parasites. They have a cell wall similar to that of other bacteria, which can be stained with special stains so that their morphology can be studied with the light microscope.

Certain arthropods, such as ticks, mites, or lice, are the natural reservoirs of rickettsiae. They are transmitted to humans by the bite of such insects, by rubbing infected insect feces into skin (for example, after a bite), or by inhaling aerosols contaminated by infected insects. The most important rickettsial pathogens are *Rickettsia prowazekii* (epidemic typhus), *Rickettsia rickettsii* (Rocky Mountain spotted fever), and *Rickettsia akari* (rickettsialpox). *Coxiella burnetii,* the etiologic agent of Q fever, has long been classified with the rickettsiae, but recent nucleic acid studies indicate they are taxonomically closer to legionellae.

Ehrlichia species are classified in the same family as rickettsiae. The two major species within the genus are *Ehrlichia chaffeensis* and *Ehrlichia ewingii*. Both organisms are transmitted to humans following the bite of an infected tick. *Ehrlichia chaffeensis* infects human blood monocytes and causes the disease human monocytic ehrlichiosis. This disease is characterized by the development of a flulike illness, with fever, headache, and muscle aches. Although the mortality rate is low (2 to 3%), more than half of infected patients require hospitalization and experience a prolonged recovery period. *Ehrlichia ewingii* causes canine granulocytic ehrlichiosis, which is a disease primarily of dogs. Humans are accidental hosts, and the clinical presentation of the disease is similar to that of *E. chaffeensis*. The patient experiences fever, headache, and muscle aches. Both infections can be treated effectively with the antibiotic tetracycline.

Following is a list of the major groups of the rickettsial family and the diseases they cause.

I. Typhus group
   A. Epidemic typhus
   B. Murine typhus
   C. Scrub typhus (tsutsugamushi fever)
II. Spotted fever group
   A. Rocky Mountain spotted fever
   B. Rickettsialpox
   C. Boutonneuse fever
III. Ehrlichiae
   A. Human monocytic ehrlichiosis
   B. Canine granulocytic ehrlichiosis

Laboratory Diagnosis

In the laboratory, rickettsiae can be propagated only in cell culture or in intact animals, such as chick embryos, mice, and guinea pigs. They are identified by their growth characteristics, by the type of injury they create in cells or animals, and by serological means. Serological diagnosis of rickettsial diseases can also be made by identifying patients' serum antibodies. Currently, PCR tests are not commercially available for diagnosing these diseases.

Chlamydiae

The chlamydiae are intermediate in size between rickettsiae and the largest viruses, which they were once thought to be. They are now recognized as true bacteria because of the structure and composition of their cell walls (the term *chlamydia* means "thick-walled") and because their basic reproductive mechanism is of the bacterial type. They are nonmotile, coccoid organisms that, like the rickettsiae, are obligate parasites. Their intracellular life is characterized by a unique developmental cycle. When first taken up by a parasitized cell, the chlamydial organism becomes enveloped within a membranous vacuole. This "elementary body" then reorganizes and enlarges, becoming what is called a "reticulate body." The latter, still within its vacuole, then begins to divide repeatedly by binary fission, producing a mass of small particles termed an "inclusion body" (see **colorplate 42**). Eventually the particles are freed from the cell, and each of the new small particles (again called elementary bodies) may then infect another cell, beginning the cycle again.

Formerly, the genus *Chlamydia* consisted of three human pathogens (*C. psittaci, C. pneumoniae*, and *C. trachomatis*). However, on the basis of taxonomic differences, two of these species, *C. psittaci* and *C. pneumoniae*, have been transferred to a new genus, *Chlamydophila*. *Chlamydophila psittaci* causes ornithosis, or psittacosis ("parrot" fever), a pneumonia transmitted to humans usually by certain pet birds. *Chlamydia trachomatis* currently is the most common bacterial agent of sexually transmitted disease; the infection often is referred to as nongonococcal urethritis. In addition, this species causes a less common sexually transmitted disease, lymphogranuloma venereum; infant pneumonitis; and trachoma, a severe eye disease that can lead to blindness. *Chlamydophila pneumoniae* produces a variety of respiratory diseases, especially in young adults. Because of difficulties growing it, the organism was identified only during the 1980s. Undoubtedly it has been causing disease for many years, if not for centuries.

Laboratory Diagnosis

Chlamydophila psittaci and *C. pneumoniae* are often diagnosed by serological means, but PCR tests are now available for diagnosing *C. pneumoniae* infections and are being used more frequently.

Mycoplasmas, Rickettsiae, Chlamydiae, Viruses, and Prions

Cell culture methods are available for growing *C. psittaci,* but isolating this organism in culture is hazardous and performed only in laboratories with specialized containment facilities.

Cell culture methods are also available for isolating *Chlamydia trachomatis,* but they are cumbersome, performed only in specialized laboratories, and generally reserved for cases of suspected child abuse. The development of nucleic acid probe and amplification assays has greatly aided diagnosis of this common sexually transmitted disease pathogen. In addition to genital specimens, eye, urine, and infant respiratory specimens may be tested, depending on the system used.

Viruses

Viruses are infectious agents that reproduce only within intact living cells. They are so small and simple in structure, and so limited in almost all activity, that they challenge our definitions of life and of living organisms. The smallest are comparable in size to a large molecule. Structurally, they are not true cells but subunits, containing only an essential nucleic acid wrapped in a protein coat, or *capsid.* The electron microscope reveals that they have various shapes, some being merely globular, others rodlike, and some with a head and tailpiece resembling a tadpole. When viruses are purified, their crystalline forms may have distinctive patterns. An intact, noncrystallized virus particle is called a *virion.*

There are many ways to classify viruses: on the basis of their chemical composition, morphology, and similar measurable properties. From the clinical point of view, it seems practical to classify them on the basis of the type of disease they produce. This, in turn, is based on their differing affinities for particular types of host cells or tissues. Thus, we speak of *neurotropic* viruses as those that have a specific affinity for cells of the nervous system. *Dermotropic* viruses affect the epithelial cells of the skin, and *viscerotropic* viruses parasitize internal organs, notably the liver. *Enteric* viruses are so called because they enter the body through the gastrointestinal tract. Their primary disease effects are exerted elsewhere, however, when they disseminate from this site of initial entry. The term *arbovirus* is used for those viruses that exist in arthropod reservoirs and are transmitted to humans by their biting insect hosts (i.e., they are *ar*thropod*bor*ne). Still other viruses, such as the human immunodeficiency virus, have effects on multiple body systems.

Ebola, previously known as Ebola hemorrhagic fever, is a rare and deadly disease caused by infection with Ebola virus. The virus was first discovered in 1976 near the Ebola River in the Democratic Republic of the Congo and since then, occasional small outbreaks of disease have occurred. In March 2014, West Africa experienced the largest outbreak of Ebola in history, with many countries affected. As of April 2016, 15,261 laboratory-confirmed cases were documented, resulting in 11,325 deaths. Due to the substandard level of medical care available in the affected countries, the actual number of cases and deaths is likely much higher. The natural reservoir for the virus is unknown, but bats are the most likely reservoir. The first human case probably resulted from the bite of an infected bat. The disease is highly contagious, and human-to-human transmission results from contact with infected body secretions (blood, feces, saliva, and semen). The Ebola virus has been detected in the semen of patients up to 9 months after they have recovered from infection. Incubation periods range from 2 to 21 days, with the onset of a multitude of symptoms consisting of high fever, severe headache, muscle pain, weakness, fatigue, profuse diarrhea, vomiting, stomach pain, and internal bleeding with bruising. Treatment consists of aggressive supportive care, including intravenous fluid replacement, balancing body electrolytes, maintaining oxygen status and blood pressure, and treating other infections that may occur. The laboratory diagnosis of infection consists of submitting appropriate specimens, such as blood, to a reference laboratory for specialized testing. In late 2016, the World Health Organization published a report on an Ebola vaccine trial, finding that the vaccine is 100% effective at preventing Ebola when given 10 or more days before exposure to the deadly virus. The vaccine is not available for sale, but the World Health Organization has collected a stockpile as a safeguard, in case another Ebola outbreak occurs.

In 2016, Zika virus emerged as an important human disease in the Western Hemisphere. Zika virus causes a mild illness, called Zika, Zika disease, or Zika fever, which is characterized by mild headache, maculopapular rash, fever, malaise, conjunctivitis with extreme light sensitivity, and joint pain. Only 20% of infected patients develop symptoms of disease, which requires no specific treatment. The disease is most commonly transmitted to humans following the bite of an infected mosquito, *Aedes aegypti,* but several other species of *Aedes* may transmit the disease. Transmission has also been documented following sexual contact with an infected male, as the virus can be present in seminal fluids; and by blood transfusion from an asymptomatic, infected donor. Zika virus is now endemic throughout Central and South America (Brazil in particular) and the Caribbean islands, especially Puerto Rico. A Zika outbreak has also occurred in south Florida, causing public health concerns that the disease may spread throughout the United States. Zika disease has attracted national attention because infected mothers are at increased risk of miscarriage or having a baby born with a small head (microcephaly) and/or with brain damage. As with Ebola, Zika infection is diagnosed by sending appropriate specimens, such as blood and urine, to a reference laboratory for specialized testing. Zika vaccine trials are in progress. In table 30.1, some important viruses are grouped in a clinical and epidemiological classification that reflects either their route of transmission or the type of disease they cause in humans.

Laboratory Diagnosis

A variety of methods may be used for the laboratory diagnosis of viral infections. These include isolation of the virus in cell culture; direct examination of clinical material to detect viral particles, antigens, or nucleic acids; cytohistological (cellular) evidence of infection; and serological assays to assess an individual's antibody response to infection. No single laboratory approach is completely reliable in diagnosing *all* viral infections. Therefore, the use of any one or a

Table 30.1 Clinical and Epidemiological Classification of Some Medically Important Viruses

Respiratory Viruses	Herpesviruses	Arboviruses (Arthropod-borne)
Influenza virus	(Dermotropic and viscerotropic)	(Viscerotropic)
Parainfluenza viruses	Varicella-zoster virus (chicken pox, shingles)	Yellow fever virus
Adenoviruses	Herpes simplex virus, types 1 and 2	Dengue fever virus
Rhinoviruses	Epstein-Barr virus (infectious mononucleosis)	Colorado tick fever virus
Respiratory syncytial virus	Cytomegalovirus	Sandfly fever virus
Mumps virus	**Exanthem Viruses**	(Neurotropic)
Hantaviruses	(Dermotropic and viscerotropic)	Eastern equine encephalitis virus
Metapneumovirus	Rubeola virus (measles)	Western equine encephalitis virus
Enteric Viruses	Rubella virus (german measles)	St. Louis encephalitis virus
Poliovirus	**CNS Virus**	Japanese B encephalitis virus
Coxsackie viruses	(Neurotropic)	West Nile virus
ECHO viruses	Rabies virus	Zika virus
Hepatitis A virus (infectious)		**Other**
Rotavirus		(Transmitted by blood)
Poxviruses		Hepatitis B virus (serum)
(Dermotropic)		Hepatitis C virus
Variola virus (smallpox)		Human immunodeficiency viruses
Cowpox virus		Ebola virus
Vaccinia virus		

Mycoplasmas, Rickettsiae, Chlamydiae, Viruses, and Prions

combination of these methods may be needed to establish a specific viral etiology of disease. The choice of method may be determined by several factors, including knowledge of the pathogenesis of the suspected viral agent, the stage of the illness, and the availability of various laboratory methods for the particular viral infection suspected. More recently, nucleic acid amplification tests, such as PCR (see Exercise 17), have replaced many of these conventional methods for the diagnosis of most viral infections.

Cell Culture

Viruses are obligate, intracellular parasites that require metabolically active cells for their replication. Most can be cultivated in mammalian cell cultures, embryonated chicken eggs, or laboratory animals, such as mice. In many clinical laboratories, cell culture has supplanted the other systems for isolating most viruses. Unfortunately, a single, universal cell culture suitable for the recovery of all viruses is not available. Because of this, several different cell culture lines are used to optimize recovery of the viral agents most common in human disease. These include Rhesus monkey kidney cells, rabbit kidney cells, human embryonic lung cells (called WI-38 cells), and human epidermoid carcinoma cells of the larynx or lung, called HEp-2 or A549 cells, respectively. These cell lines are cultivated in glass or plastic tubes or flasks using specially formulated cell culture media. The cells adhere to the glass surface and produce a confluent, single layer of growth known as a cell *monolayer* (see fig. 30.1).

The ability of a virus to infect a particular cell line depends on the presence of specific receptor sites on the cell membrane to which the virus can attach. Attachment is followed by virus entry into the cell. The presence or absence of certain receptor sites on the cell membrane surface determines the susceptibility or sensitivity of that particular cell line to viral infection.

Once a virus infects a mammalian cell, it may induce certain morphologic changes in the typical appearance of the cells, known as a cytopathic effect or CPE (see fig. 30.1). Some types of CPE caused by different viruses include generalized cell rounding, syncytia formation (fusion of cells), and plaque formation (lysis of cells). Importantly, the type of cell line infected and resultant CPE produced are extremely useful in providing the identity of the particular virus isolated. The CPE may take from 1 to 25 days to develop, depending on the virus isolated.

Certain groups of viruses, such as the influenza and parainfluenza viruses, may not produce CPE when they infect cell cultures, and thus, cell monolayers infected with them appear normal morphologically. A unique property of these viruses, however, is their ability to

Figure 30.1 Cell culture of adenovirus. The uninoculated cells on the left form an even monolayer (one cell thick) in the culture tube. Once the cells are infected with virus (right), they undergo a characteristic cytopathic effect, becoming enlarged, granular in appearance, and aggregated into irregular clusters. ©*Paul A. Granato*

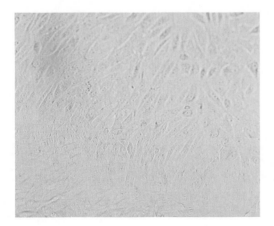
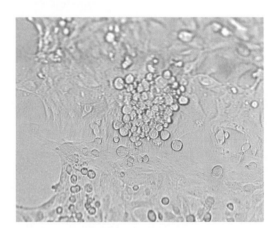

produce hemagglutinins, which are proteins projecting from the envelopes of the viruses and present in the membranes of infected cells. Hemagglutinins have the ability to adhere to erythrocytes in a process known as hemadsorption, which is used to screen certain cell cultures for the presence of influenza and parainfluenza viruses. This test is performed by overlaying the cell monolayer with a suspension of guinea pig erythrocytes, then examining for the presence of hemadsorption after 30 minutes. Adherence of the guinea pig erythrocytes to the cell monolayer is regarded as a positive test. Influenza and parainfluenza viruses are the most commonly isolated hemadsorbing viruses, but mumps virus also gives a positive reaction. Figure 30.2a is a diagrammatic representation of a negative guinea pig erythrocyte hemadsorption test and figure 30.2b illustrates a positive hemadsorption test.

Within the last several years, genetically engineered cell culture lines have been developed that are much more susceptible to viral infection than are conventional cell lines. When used in combination with specific reagents, such as monoclonal antibodies conjugated with fluorescein dyes (see Exercise 16), several common human viral pathogens are detected rapidly, often within 1 to 2 days of specimen processing and long before CPE is seen. Many clinical microbiology laboratories use genetically engineered cell culture systems for the rapid cultivation of herpes simplex virus types 1 and 2; influenza A and B viruses; parainfluenza types 1, 2, and 3 viruses; adenoviruses; respiratory syncytial virus; enteroviruses; and varicella-zoster virus from clinical specimens. The use of such rapid methods has an obvious beneficial impact on patient care and hospital infection control practices.

Despite the availability of a large number of different cell culture lines, a number of clinically important viruses cannot be grown using these conventional methods. The Epstein-Barr virus (the cause of infectious mononucleosis) and human immunodeficiency virus (the cause of AIDS) require human white blood cells for growth. Other viruses, such as some coxsackie A viruses, rabies virus, and arboviruses, are best isolated in mice. Because of the highly specialized nature of these procedures, such methods are generally performed only in reference laboratories. In addition, some viruses (e.g., hepatitis viruses and rotavirus) cannot be cultivated at all. Alternative procedures such as electron microscopy, antigen detection assays, PCR tests, or serology are used for the diagnosis of these viral infections.

Figure 30.2   Figure 30.2a is a diagrammatic representation of a negative guinea pig erythrocyte hemadsorption test. In figure 30.2b the erythrocytes adsorb to guinea pig cells infected with influenza virus.
©Paul A. Granato

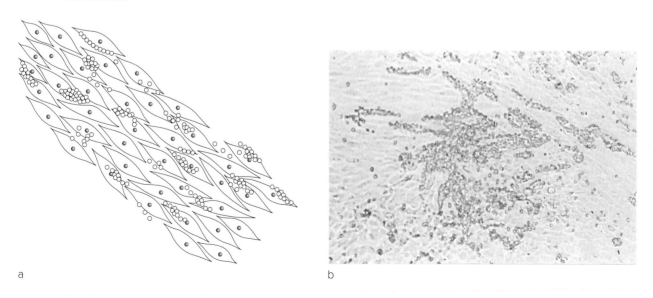

a

b

Direct Specimen Assays

Immunologic assays, such as immunofluorescence and enzyme immunoassay, are used to detect viral antigens, and nucleic acid amplification techniques are used to detect viral nucleic acids directly in patient specimens (see Exercises 16 and 17). These direct antigen detection assays are available for a large number of viruses, including respiratory syncytial virus, influenza A and B viruses, parainfluenza viruses, adenoviruses, herpes simplex types 1 and 2 viruses, varicella-zoster virus, and human metapneumovirus. Immunoassays can be used also to confirm the identity of viruses recovered in culture. **Colorplate 43** is an example of a direct immunofluorescence test, which demonstrates the presence of respiratory syncytial virus in infected human epithelial cells recovered from a nasal wash sample. **Colorplate 44** is an example of a different assay, an enzyme immunoassay, which can also detect respiratory syncytial virus from nasal wash samples. Currently, nucleic acid amplification assays are available to detect many of these viruses directly in respiratory specimens, to detect enteroviruses in cerebrospinal fluid, and to detect human papillomavirus in cervical scrapings. In addition, microarray syndrome assays, as discussed in Exercise 17, are now available to detect a large number of viral and other microbial pathogens directly in respiratory, stool, and cerebrospinal fluid specimens. Other assays are also used for quantitating levels of HIV, hepatitis C virus, and cytomegalovirus in blood. If viral products are detected, the laboratory diagnosis of infection is established, and the need to perform viral culture is eliminated. Depending on the assay, results are usually available within 10 minutes to several hours.

Cytohistological Examination

The earliest nonculture laboratory method used for viral diagnosis was screening for characteristic changes in infected human cells and tissues. Examination of cell smears or tissue sections stained with special tissue stains may reveal characteristic viral inclusion bodies that represent "footprints" of viral replication and are suggestive of certain viral infections. However, the diagnostic value of such an approach is limited because sensitivity is low (50 to 70%) compared with other available methods. The major application of this method is for the diagnosis of infections caused by viruses such as molluscum contagiosum (the cause of genital warts), which are not culturable. However, a gene amplification method is now commercially available for detecting these viruses in clinical samples.

Electron Microscopy

Electron microscopy is a powerful tool for the study of viral morphology and size but is of limited availability in most diagnostic laboratories. Direct electron microscopy also requires specimens containing high titers ($\geq 10^7$ per ml) of viral particles. The major diagnostic application of electron microscopy is for the detection of certain nonculturable viruses, particularly those that cause gastroenteritis (e.g., caliciviruses, astroviruses, and rotavirus).

Serology

Serological tests to identify a patient's antibodies are described in more detail in Exercise 33. A variety of serological tests, however, are available for the diagnosis of many viral infections. These involve the examination of two serum specimens (acute and convalescent sera spaced at least 2 to 4 weeks apart) to detect a significant change in antibody titer. Serology is extremely useful for the diagnosis of infections caused by the various hepatitis viruses.

Prions

Prion is a shorthand term for proteinaceous infectious particles. They are smaller than viruses and are believed to contain neither DNA nor RNA. Prions cause slow neurodegenerative diseases known as spongiform encephalopathies. They are classified as slow viral infections because 20 to 30 years following exposure to the agent may elapse before symptoms of infection develop in the patient. Creutzfeldt-Jakob disease and kuru are examples of human prion disease.

In the 1980s, prions attracted international scientific and public attention due to the outbreak of bovine spongiform encephalopathy, also known as "mad cow disease," in Great Britain and some other European countries. The first case in the United States was detected in 2003. Mad cow disease causes infection primarily in cattle and sheep, but human infections can result from eating infected animal meat. Prions are highly resistant to destruction and are not inactivated by thorough cooking of infected animal products. No treatments are available for diseases caused by prions, and the diseases are universally fatal, with death usually occurring within one year of the onset of symptomatic disease. In highly preliminary animal studies, researchers found that a noninfectious form of prion protein found in the brain may contribute to Alzheimer's disease, Parkinson's disease, and Lou Gehrig's disease (amyotrophic lateral sclerosis). These early-stage studies have not yet been confirmed in humans, but the findings are intriguing.

The diagnosis of prion infection is problematic. Currently, there is no clinical laboratory method available to establish the diagnosis. Instead, diagnosis is based on clinical suspicion confirmed by demonstrating characteristic spongiform changes (spongelike holes) in histological sections of brain tissue, usually postmortem. Recent evidence indicates that these spongiform changes may be seen also in more readily accessible tonsillar tissue.

Questions

1. What are mycoplasmas? How are they identified?

2. Can mycoplasmas be studied with the light microscope? If so, what kind of preparations are made?

3. What are the functions of the bacterial cell wall? How does its absence affect the behavior of bacteria?

4. How do mycoplasmas differ from other bacteria?

5. How do viruses differ from other microorganisms?

6. How are rickettsiae transmitted?

7. Name two diseases caused by *Ehrlichia* species.

8. Name the important diseases caused by *Chlamydia* and *Chlamydophila* species.

9. How are viruses identified in the laboratory?

10. What is an arbovirus?

11. Define cell culture.

12. How does the electron microscope differ from the light microscope? Describe its principles.

13. Provide a brief definition of a prion.

14. Name three human diseases caused by prions.

15. Why are prion diseases called slow viral infections?

16. Why is Zika disease considered such an important infection?

17. Complete the following table.

Disease	Type of Virus	Major Symptoms	Transmission	Immunization
Rabies				
Poliomyelitis				
Influenza				
Rubella				
Chickenpox (varicella)				
Shingles (zoster)				
Mumps				
Hepatitis A				
Hepatitis B				
Dengue				
AIDS				

18. Complete the following table.

Disease	Name of Organism	Transmission to Humans
Rocky Mountain spotted fever		
Epidemic typhus		
Rickettsialpox		
Q fever		
Trachoma		
Psittacosis		
Human monocytic ehrlichiosis		

Fungi: Yeasts and Molds

Learning Objectives

After completing this exercise, students should be able to:
1. Explain the basic structural differences between a yeast and a mold.
2. List the four major types of mycoses and give an example of each.
3. Discuss methods available for the laboratory diagnosis of a fungal infection.
4. Explain the use and importance of the germ tube test for identifying yeast.
5. Name three fungi that produce opportunistic infections.

Medical mycology is concerned with the study and identification of the pathogenic yeasts and molds, collectively called *fungi* (sing., *fungus*). You should be familiar with a number of important mycotic diseases.

Yeasts are unicellular fungi that reproduce by budding, that is, by forming and pinching off daughter cells (see **colorplate 45**). Yeast cells are much larger (about five to eight times) than bacterial cells. The best-known (and most useful) species is "bakers' yeast," *Saccharomyces cerevisiae,* used in bread making and in fermentations for wine and beer production. *Candida albicans* is the yeast that is the most common and important cause of human infection.

Molds are multicellular, higher forms of fungi. They are composed of filaments called *hyphae,* abundantly interwoven in a mat called the *mycelium.* Molds can be divided into two major groups, depending on whether they have *septate* (divided by septa) or *nonseptate hyphae.* Molds with septate hyphae produce specialized structures for reproduction called *conidiophores,* which bear numerous asexual *conidia* (sing. *conidium;* also called *spores*). The specialized reproductive structure of molds with nonseptate hyphae is called a *sporangium,* which is a saclike body that contains the spores (*sporangiospores*). When fungal spores are released by the mold, each can germinate to form new growth of the fungus. Most fungi also produce spores by a sexual process, but asexual spore production is by far their most common method of propagation. Unlike bacterial endospores, fungal spores are not highly resistant to physical and chemical antimicrobial agents. The visible growth of a mold often has a fuzzy appearance because the mycelium extends upward from its vegetative base of growth, thrusting specialized hyphae that bear sporangia or conidia into the air. This portion is called the *aerial* mycelium. You have often seen this on moldy bread or other food, and you have probably also noted that different molds vary in color (black, green, or yellow) because of their conidial pigment (see **colorplate 46**).

Most of the thousands of species of yeasts and molds that are found in nature are saprophytic and incapable of causing disease. Indeed, many are extremely useful in the processing of certain foods (such as cheeses) and as a source of antimicrobial agents. *Penicillium notatum,* for example, is the mold that produces penicillin.

Mycotic Diseases and Their Agents

Fungal diseases fall into four clinical patterns: *superficial* infections on surface epithelial structures (skin, hair, and nails), *systemic* infections of deep tissues, and *subcutaneous* and *opportunistic* infections.

Superficial Mycoses

The pathogenic fungi that cause infections of skin, hair, or nails are often referred to collectively as *dermatophytes*. These superficial mycoses are confined to the nonliving layers of the skin (the stratum corneum) and the hair and nails. They all have different names beginning with the word *tinea*, meaning worm, because of the early mistaken belief that these infections were caused by worms. The characteristic skin lesions are ringlike, and therefore, they were referred to also as ringworm, followed by the anatomic site that was affected; for example, ringworm of the scalp (tinea capitis), beard (tinea barbae), foot (tinea pedis), body (tinea corporis), groin (tinea cruris), and hand (tinea manuum). Tinea versicolor is another common superficial mycosis, but this infection is caused not by a *dermatophyte* but by a yeast called *Malassezia furfur*.

There are three major genera of dermatophytes:

Trichophyton. This genus contains many species (e.g., *T. mentagrophytes, T. rubrum,* and *T. tonsurans*) associated with "ringworm" infections of the scalp, body, nails, and feet. "Athlete's foot" is perhaps the most common of these infections.

Microsporum. There are three common species of this genus: *M. audouini, M. canis,* and *M. gypseum.* These fungi cause ringworm infections of the hair and scalp, and also of the body.

Epidermophyton. One species, *E. floccosum,* causes nail infection and ringworm of the body, including "athlete's foot." It does not infect hair.

Systemic and Subcutaneous Mycoses

Many of the fungi involved in systemic and subcutaneous infections are either yeasts or display *both* a yeast and a mold phase (they are said to be *dimorphic* because of this). The yeast phase of dimorphic fungi grows best at 35 to 37°C, whereas their mold phase grows optimally at a lower (25 to 30°C) temperature. The most important pathogenic fungi that cause systemic or subcutaneous disease are shown in table 31.1.

Table 31.1   Classification of Systemic and Subcutaneous Mycoses

Type	Sources	Entry Routes	Primary Infection	Disease	Causative Organism(s)
Primary systemic mycoses	Exogenous	Respiratory or parenteral	Pulmonary or extrapulmonary	Histoplasmosis	*Histoplasma capsulatum*
				Coccidioidomycosis	*Coccidioides immitis*
				Blastomycosis (North American)	*Blastomyces dermatitidis*
				Cryptococcosis	*Cryptococcus neoformans**
				Paracoccidioidomycosis (South American blastomycosis)	*Paracoccidioides brasiliensis*
Subcutaneous mycoses	Exogenous	Parenteral	Extrapulmonary	Sporotrichosis	*Sporothrix schenckii*
				Chromoblastomycosis	*Phialophora, Fonsecaea, Cladosporium, Rhinocladiella* species
		Skin	Subcutaneous	Mycetoma (Madura foot)	*Madurella, Pseudallescheria* species and others
Opportunistic mycoses	Endogenous	Skin, mucosae, or gastrointestinal tract	Superficial or disseminated	Candidiasis	*Candida albicans* and other *Candida* species
	Exogenous	Respiratory	Pulmonary	Aspergillosis	*Aspergillus fumigatus* and other *Aspergillus* species
	Exogenous	Respiratory or parenteral	Pulmonary or extrapulmonary	Zygomycosis	*Mucor, Rhizopus, Absidia,* and others

*Also a cause of opportunistic mycosis.

Opportunistic Mycoses

Under ordinary circumstances, fungi are of low pathogenicity and have little ability to invade the human body. However, when the host's immune defense mechanisms are decreased by illness (leukemias, lymphomas, AIDS) or by drugs (steroids, cancer chemotherapeutics, transplantation drugs), fungi (as well as other microorganisms) find the opportunity to invade and establish disease. Microorganisms that cause infections only in immunocompromised people are commonly referred to as *opportunistic pathogens*. Because fewer antimicrobial agents are available to combat fungal infections than bacterial infections, these are among the most serious opportunistic illnesses and frequently are the direct cause of the patient's death. Some opportunistic fungi, such as the yeasts *Candida* and *Cryptococcus* (see **colorplates 45** and **48**), are not always associated with immunosuppression, but others, especially species of *Aspergillus* (see **colorplates 47** and **49**) and *Mucor,* infect only immunocompromised hosts. Because the latter organisms are also widespread in the environment, health care personnel must be certain that specimens obtained from immunocompromised patients are always collected in sterile containers and in such a manner as to avoid contamination with airborne fungal spores. The microbiology technologist must also protect culture plates and broths from such contamination so that any molds that grow out are known to come from the patient and not the environment. Some agents of opportunistic fungal infections are listed in table 31.1.

Laboratory Diagnosis

The laboratory diagnosis of a fungal infection depends on the direct microscopic detection of fungal structures in clinical samples and/or the recovery in culture and subsequent identification of the fungus. Fungi may be isolated from a variety of clinical specimens representing the focus of infection (sputum, spinal fluid, tissue, pus aspirated from lymph nodes or other lesion, bone marrow aspirates, or skin scrapings). All specimens of sufficient quantity submitted for fungal culture should be examined microscopically for fungi. When there is not sufficient specimen to allow both a culture and direct microscopic examination, the culture has priority over the smear because culture is more sensitive than microscopic examination. However, observing a fungus in a clinical specimen is often valuable in establishing the significance of the fungus (i.e., ruling out contamination) and in providing early information that may be crucial for determining appropriate patient therapy.

In general, serological tests (looking for a significant change in antibody titer in paired serum specimens, see Exercise 33) have limited application for the diagnosis of most fungal infections. Exceptions to this rule include certain dimorphic fungal diseases, such as histoplasmosis and coccidioidomycosis, and aspergillosis. Nucleic acid probe technology and an exoantigen test, in which extracts of mycelial growth containing the fungal antigen are tested with known antibodies, are performed in special reference mycology laboratories. The purpose of this laboratory exercise is to acquaint the student with some direct microscopic and cultural methods that are available for establishing the laboratory diagnosis of a human mycosis.

Direct Microscopic Examinations

Wood's Lamp Exam.   The Wood's lamp is an ultraviolet light that is used to screen patients who may have superficial hair infections caused by a particular dermatophyte, or who may have a tinea versicolor skin infection. The test is performed by placing the patient in a darkened room, shining the Wood's lamp on the hair or skin, and observing for fluorescence, which is evidence of infection. A positive Wood's lamp exam of infected scalp hairs is shown in **colorplate 50**.

Histopathology.   The visualization of fungal structures (hyphae, conidia, etc.) in tissue obtained by biopsy or at autopsy *establishes* the involvement of the fungus in human disease. Specialized tissue stains such as Giemsa, methenamine silver (see **colorplate 49**), or mucicarmine may be used to facilitate the detection of the fungus in tissue. The particular fungal structures that are seen in tissue can sometimes *confirm* the identity of the fungus. For instance, in tissue, *Coccidioides immitis* typically produces spherules that contain many endospores (see fig. 31.1), whereas *Pneumocystis jiroveci* (formerly *Pneumocystis carinii*) produces a characteristic cyst form

Figure 31.1 KOH preparation of lung biopsy material showing a spherule containing endospores of *Coccidioides immitis.* Many endospores bud off from the thick-walled spherule, which has burst, releasing the endospores into the surrounding tissue. Each endospore is able to form a new spherule. In culture, this dimorphic fungus will grow as the filamentous hyphal form. ©*Josephine A. Morello*

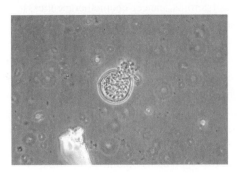

(see **colorplate 51**). Sometimes, however, the appearance of certain structures in tissue may suggest only the presence of a particular fungal group. In this latter case, culture is used to confirm the presence and identity of the fungal pathogen.

Direct Smears. Direct smears of patient material other than tissue are often made to detect the presence of fungal elements microscopically. Several types of stains or reagents are used to facilitate the detection of certain fungi:

1. Ten percent potassium hydroxide: Potassium hydroxide preparations are used to examine a variety of clinical samples including hair, nails, skin scrapings, fluids, or exudates. The potassium hydroxide solution serves to clear away tissue cells and debris, making the fungi more prominent. Slides must be examined with reduced illumination to allow fungal structures to be seen (see **colorplate 52**).

2. Calcofluor white: This reagent is used with most specimen types to detect the presence of fungi by fluorescence microscopy. The cell walls of the fungi bind the stain and fluoresce blue-white or apple green, depending on the filter combination used with the microscope. This stain is useful for examining skin scrapings for the presence of dermatophytes, and tissues and body fluids for yeasts and filamentous fungi (see **colorplate 49** [right]).

3. India ink: This traditional test is usually ordered to screen for the presence of *Cryptococcus neoformans* in spinal fluid samples. This yeast is encapsulated, and the capsule can be visualized readily against the black background of the India ink as a clear halo surrounding the yeast cell (see **colorplate 48**). The India ink test is very insensitive (detecting only 40% of cases of cryptococcal meningitis) and therefore has been superseded by other tests, such as the cryptococcal antigen latex agglutination test (same principle as in fig. 16.2), which detects more than 90% of cases of cryptococcal meningitis. For this reason, the India ink test is not performed in most clinical microbiology laboratories. More recently, nucleic acid amplification tests, such as microarray assays, have become available for detecting *C. neoformans* directly in spinal fluid specimens (see Exercise 17). The microarray assays are more sensitive than the cryptococcal antigen latex agglutination test.

4. Wright, Giemsa, or Diff-Quik stains: These specialized stains are often used on blood and bone marrow smears to look for intracellular yeast forms of *Histoplasma capsulatum.*

5. Gram stain: Most fungi are not stained well by the Gram-stain procedure, and therefore, it is of limited use when examining specimens for fungal forms. It is generally reliable only for detecting the presence of *Candida* species (see **colorplate 45**), *Sporothrix schenkii,* and perhaps a few other fungi in clinical material. In Gram-stained spinal fluid specimens, *Cryptococcus neoformans* may appear as irregularly staining gram-positive yeast cells surrounded by an orange capsule (see **colorplate 48**).

Culture

The cultural isolation of a fungus from a clinical specimen and its subsequent identification is the *definitive* test for establishing the etiology of a fungal disease. The medium most commonly used to isolate fungi from clinical specimens is Sabouraud dextrose agar. Most fungi grow well at room temperature; however, depending on the fungus, several days to weeks or months may be required for its recovery and complete characterization. In recent years, differential media have been developed to distinguish between species of yeast recovered from clinical specimens. One

Figure 31.2 Germ-tube formation by *Candida albicans*. The oval-shaped yeast cell in the center has sprouted a germ tube when incubated for 2 hours in horse serum. Not all cells in the preparation form the germ tube. The apparent halo around some cells represents light refraction and not a capsule. ©*Verna Morton*

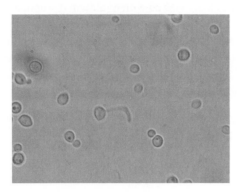

example is chromogenic agar, which differentiates species of *Candida* growing on the medium by the color produced around each colony. The use of such media allows for early genus and species identification of the isolate, which could have a favorable impact on patient care. The following discussion summarizes some of the cultural procedures used to identify yeasts and molds.

Yeasts. Yeasts, such as *Candida* species and *Cryptococcus neoformans,* are a heterogeneous group. Their identification is based on colonial and cellular morphology and biochemical characteristics. Morphology is used primarily to establish the genus identification, whereas biochemical tests are used to differentiate the various species.

1. Germ tube test: More than 90% of yeast infections are caused by *Candida albicans.* The germ tube test is a rapid and inexpensive method used to identify this species. When they are inoculated into a tube containing 1 ml of horse serum, all strains of *C. albicans* produce a specialized structure, called a germ tube, within 2 hours of incubation at 35°C (see fig. 31.2). All other yeast isolates are "germ tube negative" within that same time period, but prolonged incubation past 2 hours may result in false-positive tests.
2. Biochemical characterization: Traditional tests used for identifying yeasts to the species level involve the assimilation and/or degradation of various carbohydrates. These tests are now commercially available in kit form, much as bacterial identification systems are (see Experiment 23.6). Popular systems include the RapID yeast system, the Minitek, the API 20CAux Yeast Identification System, the Uni-Yeast-Tek System, and the automated bioMérieux Vitek 2 YST ID card, all of which are modifications of the classic carbohydrate degradation and assimilation techniques. Identification results are usually available within 24 to 72 hours.
3. Yeast may also be rapidly and reliably identified by using MALDI-TOF technology (see Exercise 18).

Molds. The identification of filamentous fungi (molds) depends on a number of factors including growth rate, colonial appearance, microscopic morphology, and some metabolic properties. A highly experienced technologist or mycologist is needed to identify most molds reliably.

1. Macroscopic appearance: A giant colony culture (a single colony grown on the center of a culture plate) is often prepared to determine the growth rate of a mold and to observe its colonial appearance (color, texture of hyphae, etc.; see **colorplate 46**). The bottom side of the plate (called the "reverse") is also examined because some fungi produce a diffusible pigment that is evident from the reverse side only. These macroscopic features are useful in the preliminary identification of the fungus.
2. Microscopic appearance: Accurate identification of a mold is based on microscopic examination of the conidia or the sporangiospores and the fungal structures on which they are borne. Microscopic preparations may be made directly from the culture (see **colorplate 53**), or mycologists may use a slide culture technique that allows these sporulating structures to be viewed microscopically at various stages of growth without disturbing their characteristic arrangements. To prepare a slide culture, a small square block of Sabouraud agar is placed on a sterile microscope slide in a sterile Petri dish. The agar is inoculated with the fungus to be identified and then covered with a cover glass. A piece of wet cotton is placed in the dish to keep the atmosphere moist and prevent drying of the agar medium. The dish and slide are incubated at room temperature or in a 25 to 30°C incubator. The slide can be viewed directly under the microscope, or the cover glass can be

removed, stained with lactophenol cotton blue, and mounted on a clean slide for viewing (see **colorplate 47**). This slide culture system improves the chances of observing fungal structures that permit genus and species identification. A Scotch tape preparation such as illustrated in figure 31.3 can be prepared from colonies growing on agar plates. This type of preparation also allows fungal structures to be viewed with minimum disruption of their characteristic morphology.

Note: In the clinical laboratory, examination of mold cultures, including the preparation of a Scotch tape mount, is always performed in a biological safety cabinet or negative pressure hood to protect the laboratory worker from aerosolized conidia (spores) and to prevent contamination of the environment. An innocuous-appearing mold growing on a culture plate inoculated with a patient specimen may be a serious, infectious pathogen.

Table 31.2 shows the characteristic morphological features of some important pathogenic fungi. In the exercise that follows you will study both fresh and prepared materials to learn these features.

Figure 31.3 Scotch tape preparation.

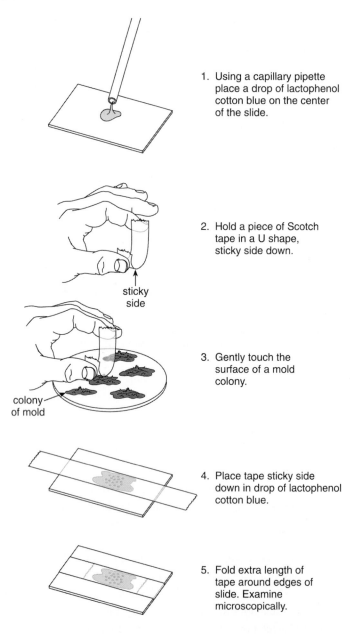

1. Using a capillary pipette place a drop of lactophenol cotton blue on the center of the slide.

2. Hold a piece of Scotch tape in a U shape, sticky side down.

sticky side

3. Gently touch the surface of a mold colony.

colony of mold

4. Place tape sticky side down in drop of lactophenol cotton blue.

5. Fold extra length of tape around edges of slide. Examine microscopically.

Microbial Pathogens Requiring Special Laboratory Techniques

Table 31.2 Some Important Pathogenic Fungi

Organisms	Morphological Features	Diseases
Yeasts or yeastlike		
Cryptococcus neoformans	Yeasty soft colonies	Pneumonia, meningitis, other tissue infections
	Encapsulated budding cells	
Candida albicans	Budding cells, pseudomycelium, and chlamydospores	Skin and mucosal infections, sometimes systemic
Systemic fungi		
Histoplasma capsulatum	*In tissues,* intracellular and yeastlike	Histoplasmosis is primarily a disease of the lungs; may progress through the mononuclear phago-cyte system to other organs
	In culture at 37°C, a yeast	
	In culture at room temperature, a mold with character-istic macroconidia (spores)	
Coccidioides immitis	*In tissues,* produces spherules filled with endospores	Coccidioidomycosis is usually a respiratory disease; may become disseminated and progressive
	In culture, a cottony mold with fragmenting mycelium	
Blastomyces dermatitidis	*In tissues,* a large thick-walled budding yeast	North American blastomycosis is an infection that may involve lungs, skin, or bones
	In culture at 37°C, a yeast	
	In culture at room temperature, a mold	
Paracoccidioides brasiliensis	*In tissues,* a large yeast showing multiple budding	Paracoccidioidomycosis (South American blastomycosis) is a pulmonary disease that may become disseminated to mucocutaneous membranes, lymph nodes, or skin
	In culture at 37°C, a multiple budding yeast	
	In culture at room temperature, a mold	
Subcutaneous fungi		
Sporothrix schenckii	*In tissues,* a small gram-positive, spindle-shaped yeast	Sporotrichosis is a local infection of injured subcutaneous tissues and regional lymph nodes
	In culture at 37°C, a yeast	
	In culture at room temperature, a mold with character-istic spores	
Cladosporium Fonsecaea Phialophora }	*In tissues,* dark, thick-walled septate bodies	Chromoblastomycosis is an infection of skin and lymphatics of the extremities caused by any one of several species
	In culture, darkly pigmented molds	
Madurella Pseudallescheria} Curvularia and others }	*Tissue* and *culture* forms vary with causative fungus	Mycetoma (maduromycosis, madura foot) is an infection of subcutaneous tissues, usually of the foot, caused by any one of several species
Superficial fungi		
Microsporum species Trichophyton species Epidermophyton floccosum	These fungi grow in cultures incubated at room tem-peratures as molds, distinguished by the morphol-ogy of their reproductive spores	Ringworm of the scalp, body, feet, or nails

Purpose	To observe the microscopic structures of some fungi
Materials	Sabouraud agar slant culture of *Candida albicans*
	Tubes containing 1.0 ml of inactivated horse serum
	Sabouraud agar plate cultures of *Aspergillus, Rhizopus, Penicillium*
	Blood agar plates exposed 3 to 5 days earlier for 30 minutes at home, in class, public transportation, etc.
	Glass microscope slides and coverslips
	Transparent tape (e.g., Scotch tape)
	Dropper bottles containing lactophenol cotton blue
	Capillary pipettes and pipette bulbs
	Prepared slides of dermatophytes
	Prepared slides of yeast and mold phases of a systemic fungus
	Projection slides, if available
	Marking pen or pencil

Procedures

1. Pick up a small amount of yeast growth from the tube of *Candida albicans.*
2. *Lightly* inoculate a tube of horse serum with the growth. Do not make a turbid suspension. Incubate the tube at 35°C for 2 hours.
3. At the end of 2 hours, use a capillary pipette to place a small drop of the serum suspension on a microscope slide. Cover the drop with a coverslip. Examine the slide under the low and high-dry power of your microscope. You will need to reduce the light intensity by partially closing the iris diaphragm of your microscope (see fig. 1.1).
4. If the test is positive, you should see a small stalk, or "germ tube," sprouting from several of the yeast cells. This confirms the identification of the yeast as *Candida albicans.*
5. Make a drawing of the yeast cells that you see with and without germ tubes.
6. Make a Gram stain of the *Candida* culture and make drawings of your observations.
7. Prepare a transparent tape preparation of the *Aspergillus, Rhizopus,* and *Penicillium* growth as follows (see fig. 31.3): Place a drop of lactophenol cotton blue on a clean microscope slide. Carefully uncover a plate culture of one of the molds. Cut a piece of transparent tape slightly longer than the length of the microscope slide (about 4 inches). Hold the tape in a ∪ shape with the sticky side down and *gently* touch the surface of a mold colony. Some of the colony growth will adhere to the tape. Place the tape with the sticky side down across the microscope slide so that the colony growth is in contact with the lactophenol cotton blue. Fold the extra length of tape around the edges of the slide. Examine the slide under the low and high-dry objectives of your microscope, as you did the germ tube preparation.
8. Repeat the procedure with any molds you see growing on the blood agar plates that you exposed to the environment.
9. Record your results.
10. Examine the prepared slides and make drawings of your observations.

Results

1. Draw a diagram showing all the structures of *Candida albicans* that you observed.

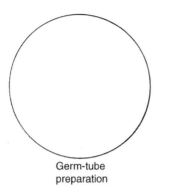

Germ-tube
preparation

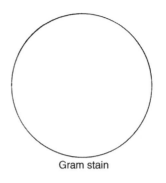

Gram stain

Gram reaction_____

2. List the principal differences you have observed in yeast cells as compared with bacteria.

3. Draw the spores, spore-bearing structures, and hyphae of each of the following:

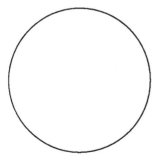

Aspergillus

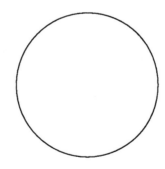

Rhizopus

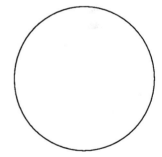

Penicillium

Colony color? _____   _____   _____

4. Draw the spores, spore-bearing structures, and hyphae of three molds growing on the blood agar plates you exposed to the environment. Do they resemble any of the fungi you observed previously?

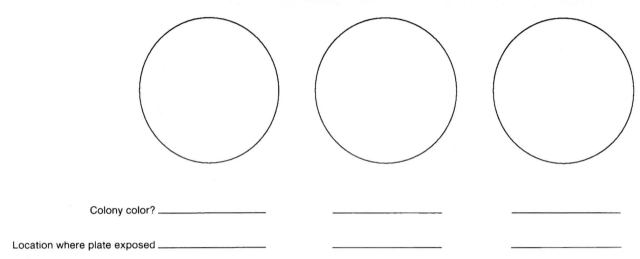

Colony color? _____   _____   _____

Location where plate exposed _____   _____   _____

5. From the prepared or projection slides, draw the microscopic structures you have seen in each phase of a systemic fungus.

Questions

1. For each of the diseases listed, indicate the type of specimen(s) that should be collected for laboratory diagnosis.

   Cryptococcosis: _____

   Athlete's foot: _____

   Tinea capitis: _____

   Candidiasis: _____

   Histoplasmosis: _____

2. What is a superficial mycosis?

3. How would you recognize a patient's ringworm infection? Would you take any special precautions in collecting a clinical sample of the ringworm lesion? If so, explain.

4. Should hospitalized patients who share the use of a shower room wear protective slippers when using it? Why?

5. What are some of the valuable uses of saprophytic fungi?

6. How is the Wood's lamp used in the diagnosis of tinea capitis?

7. From what source do patients with *Aspergillus* infections acquire the organism?

8. What is the advantage of viewing mold structures in a transparent tape preparation?

9. What fungus can be identified reliably by using the germ tube test?

10. Name three stains or reagents that may be used to facilitate the microscopic detection of fungi in clinical samples.

11. What is the main advantage of using the slide culture technique for identifying molds?

12. When working with molds in the laboratory, why should all preparations be performed under a hood or in a biological safety cabinet?

Protozoa and Animal Parasites

Learning Objectives

After completing this exercise, students should be able to:
1. Describe three basic cellular structures found in all protozoa.
2. Name the six major groups of protozoa and give an example of each.
3. Define the term helminth and distinguish a roundworm from a flatworm.
4. Compare extraintestinal and intestinal parasitic infections and provide two examples of each.
5. List three methods for establishing a laboratory diagnosis of an intestinal parasitic infection.

Medical parasitology is concerned with the study and identification of the pathogenic protozoa and helminths (worms) that cause the parasitic diseases of humans and animals.

Protozoa

Protozoa are the largest of the unicellular microorganisms. They are classified in the kingdom *Protista* although their name implies that they were the forerunners of the animal kingdom (*proto* = first; *zoa* = animals).

The basic structures of all protozoa include a *nucleus* well defined by a *nuclear membrane,* lying within *cytoplasm* that is enclosed by a thin outer *cell membrane.* Other specialized structures, such as cilia or flagella (see **colorplate 54**) for locomotion or a gullet for food intake, vary with different types of protozoa. Six major groups of protozoa are distinguished on the basis of their locomotory structures or their reproductive mechanisms.

Amebae. Simple *ameboid* forms. Move by bulging and retracting their cytoplasm in any direction. Major pathogen is *Entamoeba histolytica* (see **colorplate 55**).

Ciliates. Move by rapid beating of *cilia* (fine hairs) that cover the cell membrane. *Balantidium coli* is a protozoan ciliate that may cause human disease.

Flagellates. Possess one or more *flagella* that give them a lashing motility. *Giardia lamblia,* also known as *Giardia duodenalis* (see **colorplate 56**), *Trichomonas vaginalis* (see **colorplate 54**), and the trypanosomes are the major pathogens in this group.

Apicomplexa. No special structures for locomotion (some immature forms have ameboid motility). Reproductive cycle includes both immature and mature forms (later called *sporozoites*). *Toxoplasma gondii* and *Plasmodium* species (see **colorplate 57**), which are the malarial parasites, are the representative pathogens in this group.

Coccidia. Represent a subphylum of the Apicomplexa. Coccidia have a complex life cycle in which all stages of parasite development are intracellular. Major genera include *Cryptosporidium* (see **colorplate 58**), *Cyclospora,* and *Isospora.*

Microspora. Includes a large group of obligate, intracellular protozoa that produce spores. These protozoa are classified in more than 100 genera and 1,200 species, collectively called *microsporidia.* Major genera causing human disease are *Enterocytozoon, Encephalitozoon, Nosema,* and *Pleistophora.*

Diagrammatic examples of the amebae, ciliates, flagellates, and the apicomplexans are shown in fig. 32.1.

As indicated, species from each of these protozoan groups are associated with human diseases. Some of them are carried into the body through the gastrointestinal tract (in

Figure 32.1 Diagrams of four types of protozoa. (a) An active ameba. (b) A ciliated protozoan (*Balantidium coli*). (c), (d), and (e) Three types of flagellated protozoa. (f) Developmental stages of the malarial parasite, a sporozoan (*Plasmodium* species).

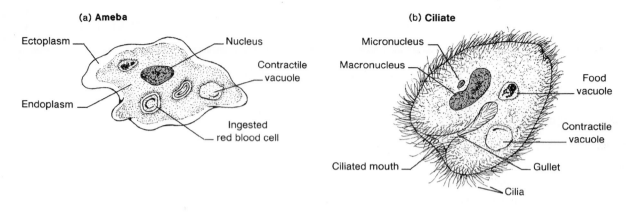

(a) **Ameba**

Ectoplasm
Nucleus
Contractile vacuole
Endoplasm
Ingested red blood cell

(b) **Ciliate**

Micronucleus
Macronucleus
Food vacuole
Contractile vacuole
Ciliated mouth
Gullet
Cilia

FLAGELLATED PROTOZOA

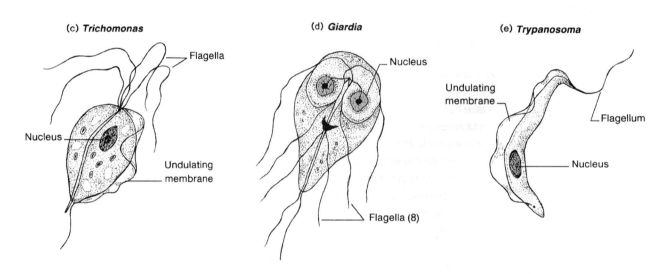

(c) *Trichomonas*

Flagella
Nucleus
Undulating membrane

(d) *Giardia*

Nucleus
Flagella (8)

(e) *Trypanosoma*

Undulating membrane
Flagellum
Nucleus

(f) **A sporozoan. The malarial parasite's life cycle**

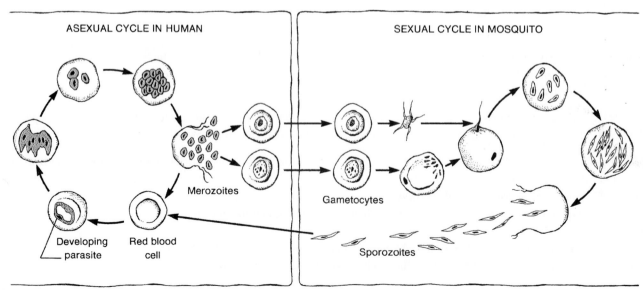

ASEXUAL CYCLE IN HUMAN

SEXUAL CYCLE IN MOSQUITO

Merozoites
Developing parasite
Red blood cell
Gametocytes
Sporozoites

Table 32.1   Pathogenic Protozoa

Disease	Type of Protozoa	Name of Organism	Entry Route
Amebiasis (dysentery)	Ameba	*Entamoeba histolytica*	Ingestion
Balantidiasis (dysentery)	Ciliate	*Balantidium coli*	Ingestion
Giardiasis (diarrhea)	Flagellate	*Giardia lamblia (Giardia duodenalis)*	Ingestion
Trichomoniasis (vaginitis) (see **colorplate 54**)	Flagellate	*Trichomonas vaginalis*	Sexual transmission
Trypanosomiasis: African sleeping sickness American form: Chagas disease	Flagellate	*Trypanosoma brucei gambiense* *T. brucei rhodesiense* *T. cruzi*	Arthropod bite Arthropod bite
Leishmaniasis: Kala-azar American form: espundia	Flagellate	*Leishmania donovani* *L. braziliensis* *L. mexicana*	Arthropod bite Arthropod bite
Malaria (see **colorplate 57**)	Apicomplexan	*Plasmodium vivax* *P. malariae* *P. falciparum*	Arthropod bite
Toxoplasmosis (systemic infection)	Apicomplexan	*Toxoplasma gondii*	Ingestion or congenital
Cryptosporidiosis (diarrhea)	Coccidian	*Cryptosporidium parvum*	Ingestion
Microsporidiosis (diarrhea, systemic)	Microsporidian	*Enterocytozoon bieneusi* and others	Direct contact Ingestion Inhalation

contaminated food or water, or by direct fecal contamination of objects placed in the mouth), localize there, and produce diarrhea or dysentery. Others are carried by arthropods, which inject them into the body when they bite. This group of protozoa then infects the blood and other deep tissues. The pathogenic protozoa are summarized in table 32.1.

It should also be noted that some of the intestinal protozoa may live normally in the bowel without causing damage under ordinary circumstances. Some flagellated protozoa frequently are found on the superficial urogenital membranes and sometimes are troublesome when they multiply extensively and irritate local tissues.

Other amebae live freely in the environment, in soil and water. Under special circumstances, some of these organisms can infect humans. Members of the genus *Naegleria* inhabit freshwater ponds, lakes, and quarries. When people dive or swim in water containing the amebae, the organisms can be forced up with water, through the thin nasal passages, directly into the central nervous system to cause an almost universally fatal meningoencephalitis (affects both meninges and brain). *Acanthamoeba* species (see fig. 32.2) are associated with corneal infections in persons whose contact lenses or contact lens care solutions become contaminated by the amebae. To avoid infection, these lenses and lens care solutions must be kept meticulously clean. Corneal transplant is usually required for patients with *Acanthamoeba* eye infection.

Parasitic Helminths

Helminths, or worms, are soft-bodied invertebrate animals. Their adult forms range in size from a few millimeters to a meter or more in length, but their immature stages (eggs, or *ova,* and

Figure 32.2 An *Acanthamoeba* trophozoite (bottom center) and cysts (refractile objects at top right) isolated from the contact lens of a patient who required a corneal transplant because of the infection. The tiny objects throughout the background are cells of *Escherichia coli* on which the amebae feed when grown in culture. *©Josephine A. Morello*

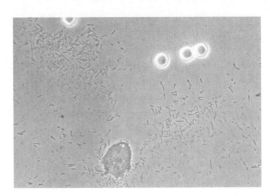

larvae) are of microscopic dimensions. Relatively few species of helminths are parasitic for humans, but these few are widely distributed. It has been estimated that 30% of the earth's human inhabitants harbor some species of parasitic worm.

There are two major groups of helminths: the *roundworms,* or nematodes, and the *flatworms,* or platyhelminths. The latter are again subdivided into two groups: the *tapeworms* (cestodes) and *flukes* (trematodes). A summary of the major characteristics of these groups is given here:

Roundworms (Nematodes).   Roundworms are cylindrical worms with bilateral symmetry. Most species have two sexes, the female being a copious egg producer. These ova hatch into larval forms that go through several stages and finally develop into adults. In some instances, the eggs of these worms are infective for humans when swallowed. In the intestinal tract, they develop into adults and produce local symptoms of disease. In other cases, the larval form, which develops in soil, is infective when it penetrates the skin and is carried through the body, finding its way finally into the intestinal tract where the adults develop. In the case of *Trichinella* (the agent of trichinosis), the larvae are ingested in infected meat, but penetrate beyond the bowel and become encysted in muscle tissue. One group of roundworms, the *filaria,* e.g., *Wuchereria* spp., are carried by arthropods and enter the body by way of an insect bite (see table 32.2).

Flatworms (Platyhelminths).   Flatworms are flattened worms that also show bilateral symmetry. Some are long and segmented (tapeworms); others are short and nonsegmented. Most are hermaphroditic (i.e., they contain both male and female reproductive organs).

Tapeworms (Cestodes).   Tapeworms are long, ribbonlike flatworms composed of individual segments (*proglottids*), each of which contains both male and female sex organs. The tiny head, or *scolex,* may be equipped with hooklets and suckers for attachment to the intestinal wall. The whole length of the tapeworm, the *strobila,* may have only three or four proglottids or several hundred. Eggs are produced in the proglottids (which are then said to be *gravid*) and are extruded into the bowel lumen. Often the gravid proglottids break away intact and are passed in the feces. All tapeworm infections are acquired through ingestion of an infective immature form, in most cases larvae encysted in animal meat or fish (e.g., *Diphyllobothrium latum;* see **colorplate 59**). Usually development into adult forms occurs in the intestinal tract, and the tapeworm remains localized there. In one type of tapeworm infection, echinococcosis, the eggs are ingested, penetrate out of the bowel, and develop into larval forms in other tissues (see **colorplate 60**).

Flukes (Trematodes).   Some flukes are short, ovoid or leaf-shaped, and hermaphroditic; others are elongate, thin, and bisexual. The flukes are not segmented. They are usually grouped according to the site of the body where the adult lives and produces its eggs, that is, blood, intestinal, liver, and lung flukes. Some of these infections are acquired through the ingestion of larval forms encysted in plant, fish, or animal tissues. In others, a larval form (swimming freely in contaminated water) penetrates the skin and makes its way into deep tissues.

Table 32.2 summarizes the important helminths that cause disease in humans.

Table 32.2   Important Helminths of Humans

Parasite	Transmission	Entry Route
Roundworms		
Enterobius vermicularis (pinworm)	Eggs, via direct fecal contamination	Mouth
Trichuris trichiura (whipworm)	Eggs matured in soil	Mouth
Ascaris lumbricoides	Eggs matured in soil	Mouth
Necator americanus (hookworm)	Larvae matured in soil	Skin
Trichinella spiralis	Larvae in infected pork or other animal tissue	Mouth
Wuchereria and others (filarial worms)	Larvae in arthropod host	Skin
Tapeworms		
Taenia solium (pork tapeworm)	Larvae in infected pork	Mouth
Taenia saginata (beef tapeworm)	Larvae in infected beef	Mouth
Diphyllobothrium latum (fish tapeworm)	Larvae in infected fish	Mouth
Echinococcus granulosus	Eggs in dog feces	Mouth
Blood flukes		
Schistosoma species	Larvae swimming in water	Skin (or mucosa)
Liver fluke		
Clonorchis sinensis	Larvae in marine plants or fish	Mouth
Lung fluke		
Paragonimus westermani	Larvae in infected crustaceans	Mouth
Intestinal fluke		
Fasciolopsis buski	Larvae in marine plants or fish	Mouth

Laboratory Diagnosis

Almost all parasitic diseases, whether intestinal or extraintestinal, are diagnosed by finding the organism in appropriate clinical specimens, usually by microscopic examination. Intestinal infections are generally limited to the bowel, and therefore, fecal material is the specimen of choice. In extraintestinal infections, the diagnostic stage of the parasite may be found in blood, tissue, or exudates, so that these specimen types must be examined. With rare exceptions, such as extraintestinal amebiasis and toxoplasmosis, routine serological tests have no application in the diagnosis of parasitic diseases.

Intestinal Parasitic Infections

Protozoa or helminths may cause intestinal parasite infections. The laboratory diagnosis of these diseases depends almost exclusively on finding the diagnostic stage(s) in fecal material. Alternatively, enzyme immunoassay, direct fluorescence antibody tests, or microarray assays may be used for detecting the presence of protozoa, such as *Cryptosporidium parvum* (see **colorplate 58** right), *Giardia lamblia* (*duodenalis*), and *Entamoeba histolytica,* in stool samples. If stool samples cannot be examined immediately after passage, a portion of the stool must be placed in a stool collection kit with a special preservative to maintain the structural integrity and morphology of the diagnostic cysts, eggs, or larvae. There is no one perfect stool preservative, and the choice usually depends on the laboratory that performs the analysis.

Once a stool is received by the laboratory, the ova and parasite (O&P) examination may consist of any combination or all three of the following techniques: direct wet mount, concentration, and permanent stained smear. Each technique is designed for a particular purpose. Traditionally, the direct examination is used to detect protozoan motility. Since most laboratories use a stool preservative that kills protozoa, direct wet-mount examinations for this purpose are not routinely performed. Instead, the direct wet-mount exam may be used to screen for cysts and eggs that may be present in large numbers in the fecal sample.

Fecal concentration procedures allow for the detection of small numbers of organisms that may be missed when only a direct smear is examined. There are two types of concentration procedures: sedimentation and flotation. Both are designed to separate protozoan cysts and oocysts, microsporidian spores, and helminth eggs and larvae from fecal debris by centrifugation (sedimentation) or differences in specific gravity (flotation).

Stained smears can also be prepared from fecal samples to allow for the improved detection and identification of intestinal protozoa. These slides serve as a permanent record of the organism identified and may be used for teaching purposes as well. Three stains commonly used for the detection of intestinal parasites are the trichrome, iron-hematoxylin, and modified acid-fast stains.

Microscopic examination of fecal material requires the eye of a skilled and experienced microbiologist for the reliable detection of intestinal parasites. Fecal samples often contain artifact material such as fibers, vegetable cells, crystals, etc., which can easily be mistaken for the diagnostic structures of intestinal parasites.

Intestinal Protozoa.   The protozoa that parasitize the human intestinal and urogenital systems belong to five major groups: amebae, flagellates, ciliates, coccidia, and microsporidia. With the exception of the flagellate *Trichomonas vaginalis* (an important cause of vaginitis, see **colorplate 54**) and microsporidia of the genera *Pleistophora, Nosema,* and *Encephalitozoon,* all of these organisms live in and may cause disease of the intestinal tract.

Intestinal Helminths.   Intestinal helminths are usually diagnosed by the microscopic detection of their eggs or larvae in feces. Characteristics used in identification include size, shape, thickness of shell, special structures of the shell (mammillated covering, operculum, spine, or knob), and the developmental stage of egg contents (undeveloped, developing, or embryonated). Figure 32.3 shows the relative sizes and comparative morphologies of representative helminth eggs.

Extraintestinal Parasitic Infections

Blood and Tissue Protozoa.   Among the protozoa that parasitize human blood and tissue, malaria is detected most frequently in the United States. The laboratory diagnosis of malaria is made by examining blood smears collected from the patient. Blood smears are stained with Giemsa or Wright stain, the common stains also used to examine blood films for hematological studies. These stains help distinguish the various diagnostic stages and allow for the identification of *Plasmodium* species (see **colorplate 57**). Of the four human malarial parasites, *Plasmodium vivax* and *Plasmodium falciparum* account for more than 95% of infections, with *P. vivax* responsible for about 80% of these. Identification of malarial parasites to the species level is important for establishing the prognosis of the disease and predicting the likelihood of drug resistance. Many strains of *P. falciparum*, the species that causes a more serious, life-threatening form of the disease, are now resistant to chloroquine, the drug of choice for treatment. Other more exotic and far less common blood and tissue protozoan diseases seen in the United States are leishmaniasis and trypanosomiasis. These infections, as well as malaria, are almost universally imported into the United States by persons arriving from countries where the parasitic agents are endemic.

Toxoplasma gondii is a tissue protozoan that is an established cause of congenital disease. More recently, toxoplasmosis has been recognized as a cause of central nervous system disease in HIV-infected patients. The diagnosis of toxoplasmosis often depends on the detection of the organism in tissue biopsy material, CSF specimens, or buffy coat of blood (the white blood cell layer that forms between the erythrocytes and plasma when anticoagulated blood is lightly centrifuged). In general, however, such specimens do not reveal the parasites, even in the presence of active disease. Therefore, serological tests are recommended in all suspected cases of toxoplasmosis.

Figure 32.3 Relative sizes and comparative morphologies of representative helminth eggs. Modified from Centers for Disease Control and Prevention. Source: Modified from Centers for Disease Control and Prevention.

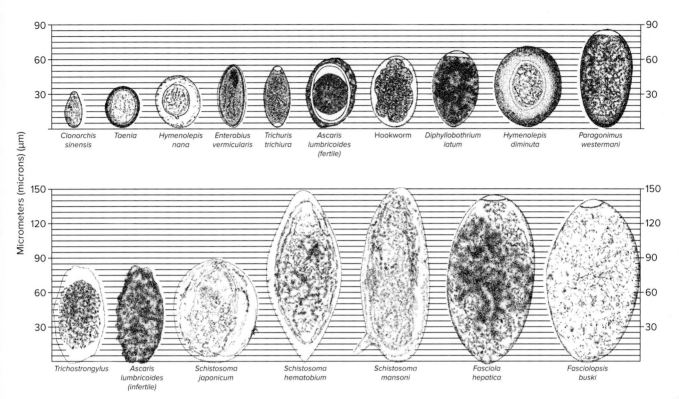

Tissue Helminths. A large number of helminthic parasites, including nematodes, flukes, and tapeworms, live in human tissues as adults or larvae. Diagnosis of infections caused by them often depends on the identification of the parasite's reproductive products discharged in blood, feces, or other body fluids or, in the case of larval parasites, on the recovery from or detection of the parasite itself in tissue.

Some of the more common tissue helminths are listed here for your review. The nematodes include the filarial worms *Wuchereria bancrofti, Brugia malayi, Onchocerca volvulus,* and *Loa loa. Strongyloides* species (a cause of cutaneous larva migrans), *Toxocara canis* (the cause of visceral larva migrans), and *Trichinella spiralis* (the cause of trichinosis) are nematodes that cause disease in the United States. The trematodes include the liver flukes (*Fasciola hepatica, Clonorchis sinensis,* and *Opisthorchis viverrini*), lung flukes (*Paragonimus westermani*), and the blood flukes (*Schistosoma* species). Finally, there are the cestodes or tapeworms, some of the more common of which include *Taenia solium* (the cause of cysticercosis), *Diphyllobothrium latum* (the fish tapeworm, see **colorplate 59**), and *Echinococcus granulosus* and *Echinococcus multilocularis* (the causes of hydatid cyst disease, see **colorplate 60**).

Except where noted, people with tissue helminth diseases become infected outside of the United States. Because of the current ease and frequency of global travel, however, microbiologists throughout the world must become familiar with the laboratory diagnosis of these infections.

Prepared slides and demonstration material will be studied in this exercise.

Purpose	To study the microscopic morphology of some protozoa and parasitic helminths, and to learn how parasitic diseases are diagnosed
Materials	Prepared slides of protozoa
	Prepared slides of helminth adults, eggs, larvae
	Projection or PowerPoint slides, if available

Procedures

1. Examine the prepared slides and audiovisual or reading material, and make drawings of different forms of protozoa and helminths.
2. Review demonstration material and assigned reading on the transmission and localization of parasites, and complete the table provided under Questions.

Results

Draw each type of organism listed:

An ameba:

A ciliated protozoan:

A flagellated protozoan:

A protozoan found in blood:

An adult roundworm:

An adult tapeworm:

A helminth egg:

Name _____ Class _____ Date _____

Questions

1. Complete the following table.

Parasite	Localization in Body (For Helminths, the Adult Form)	Specimens for Laboratory Diagnosis
Entamoeba histolytica		
Trichomonas vaginalis		
Trypanosoma brucei gambiense		
Plasmodium vivax		
Toxoplasma gondii		
Naegleria fowleri		
Enterobius vermicularis		
Ascaris lumbricoides		
Necator americanus		
Trichinella spiralis		
Taenia saginata		
Echinococcus granulosus		
Schistosoma		
Clonorchis sinensis		

2. Describe the basic structures of protozoa. Can these same structures be seen in bacteria using the light microscope?

3. Are any parasitic diseases directly communicable from person to person? If so, how are they transmitted? What kinds of precautions should be taken in caring for patients with directly transmissible parasitic infections?

4. What is an arthropod? How can it transmit infection to humans?

5. What parasitic forms can be seen in the feces of a patient with hookworm? Cryptosporidiosis? Tapeworm? Trichinosis? Malaria?

6. What parasitic forms can be seen in the blood of a patient with African sleeping sickness? Filariasis? Amebiasis?

7. What is meant by the "life cycle" of a parasite? What importance does it have to those who take care of patients with parasitic diseases?

8. What precautions should be taken to prevent infection by "free-living" amebae?

Serodiagnosis of Infectious Disease

Learning Objectives

After completing this exercise, students should be able to:
1. Discuss the two major classes of antibodies that are produced to combat infection.
2. Define briefly an anamnestic response.
3. Explain the importance of acute and convalescent serum specimens for the serologic diagnosis of infection.
4. Name four types of serologic tests and give an example of each.
5. Explain the importance of titer in making a serological diagnosis.

When an individual is infected with a microorganism, the immune system produces antibodies to help combat infection. If this is the first time that the individual has been infected with the microorganism, the antibody response is called the primary immune response. As shown in figure 33.1, two major classes of antibodies, which are found circulating in the bloodstream, are produced during the primary immune response. The first class of antibody produced is immunoglobulin M, or IgM, followed by the production of the second class of antibody, immunoglobulin G, or IgG. If the individual does not become infected with the same microorganism again, the levels of circulating IgM and IgG antibodies gradually decline. On occasion, the person may be infected again with the same microorganism. This reinfection results in a secondary antibody response, often called the memory, or *anamnestic, response*, with the production mainly of IgG antibody and very small amounts of IgM. The production of these

Figure 33.1 The primary and secondary antibody responses to an immunogenic stimulus.

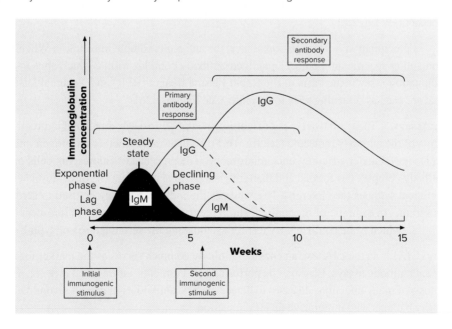

antibodies has practical value because their detection in a patient's serum may indicate recent or past infection with a particular microorganism. This is the underlying basis of serological testing, or the *serodiagnosis* of infectious disease. The branch of immunology that specifically deals with such testing is called *serology*. The tests performed in the serology laboratory not only detect the presence of IgM and/or IgG antibodies in serum, but in most cases, can measure or quantify the amount of antibody present.

One way the serodiagnosis of an infectious disease can be established is by collecting serum samples at two time periods: during the acute and the convalescent stages of the patient's disease. Ideally, the *acute* sample is collected during the symptomatic stage of disease and the *convalescent* sample is collected a week or two later when the patient is recovering. Together, acute and convalescent serum specimens are referred to simply as *paired* serum specimens. Paired samples are tested at the same time in the serology laboratory for the presence of specific antibodies against the particular microbial agent suspected to have caused the patient's illness. As seen in figure 33.1, the level or concentration of antibody increases or decreases over several weeks. If antibodies are detected, they can also be quantified by laboratory tests to determine the level of antibody present, or *titer* of the serum (to be described later in this exercise). The serodiagnosis of infection is considered established when a significant (usually fourfold) increase or decrease in antibody levels is detected between the acute and convalescent serum samples. Serological testing can also be performed on a single serum sample, in which case demonstrating a high level of antibody against a microbial agent is evidence of recent infection. Whenever possible, testing paired serum samples is preferred for diagnosis, although serodiagnosis can be achieved by detecting specific IgM antibodies in a patient's serum, as will be discussed in the section on enzyme immunoassay. The advantages of this method are that it eliminates the need for collecting acute and convalescent sera, and the test results are usually available in a shorter time.

A variety of serologic tests for detecting antibody in serum have been developed. Some of the more frequently used assays are reviewed here briefly and have been described in greater detail in Exercise 16.

Agglutination Assays. In agglutination assays, a specific antigen is attached either to a small diameter (1 μm) latex particle or to erythrocytes, and a small volume of serum is mixed with the particles. If antibody is present in the serum, an agglutination reaction occurs, as visualized by clumping of the latex particles or erythrocytes (in this latter case, also referred to as hemagglutination). A latex agglutination test is commonly used for the detection of cryptococcal antigen in spinal fluid and typing of groups A and B streptococci (see fig. 20.1).

Precipitation Assays. In precipitation or flocculation assays, the antigen is soluble in solution. When the antigen solution is mixed with a serum containing specific antibodies, an antigen-antibody complex forms, which appears as a precipitate in the reaction tube. The precipitation assay is the basis of the Rapid Plasma Reagin (RPR) and Venereal Disease Research Laboratory (VDRL) assays that are used as screening tests for syphilis (see Exercise 26).

Fluorescent Antibody Assays. There are two types of fluorescent antibody assays, direct and indirect, in which a fluorescein dye is attached to a region of the antibody molecule (see fig. 16.1). To determine whether an antigen-antibody reaction has occurred, the preparation is viewed under a fluorescence microscope. The presence of fluorescing cells or other structures indicate that an antigen-antibody complex has formed and therefore is a positive reaction. The direct test is used primarily to detect microbial antigens in clinical samples (see Exercise 16), whereas the indirect test may be used to screen for the presence of antibody in a patient's serum. An example of the indirect fluorescent antibody test is the Fluorescent Treponemal Antibody Absorption (FTA-ABS) test, which is the confirmatory test for establishing the serodiagnosis of syphilis (see Exercise 26).

Enzyme Immunoassay (EIA). In the EIA test, an enzyme-antibody complex serves as the marker for antigen-antibody reactions. Like the fluorescent antibody assays, EIA may be performed as a direct or indirect test (see fig. 16.3). The indirect EIA is used in the serodiagnosis of many infectious diseases to detect specific antibodies in patient serum. One example is its use as a screening test to identify patients who might have an HIV infection.

The EIA is also used to identify the IgM and IgG classes of antibody. As described earlier, during most microbial infections these two major antibody classes are produced by the immune system and secreted into the blood (see fig. 33.1). This protective antibody response is called *humoral immunity*. Immunoglobulin M, or IgM, is produced first, followed shortly by the production of immunoglobulin G, or IgG. IgM antibody appears early in the course of disease, decreasing to undetectable levels after several weeks. In contrast, IgG antibody appears later and remains elevated for a much longer time. EIA tests are available to specifically screen for the presence of IgM and IgG antibodies. The detection of IgM antibodies against a particular microbial agent is serologic evidence of recent infection, whereas detection of IgG antibodies only is considered evidence of an infection that occurred sometime in the past.

Once an antibody is found in a serum sample, the serology laboratory may need to quantify the level of antibody present, that is, the antibody *titer*. The titer is determined by serially diluting the patient's serum sample in test tubes or a microtiter tray. Each tube or well is then assayed for antibody by the method that was first used to detect it (e.g., agglutination, precipitation, EIA). Titer is defined as the reciprocal of the highest dilution of patient serum that produces a detectable reaction. For example if a positive reaction is seen in a serum sample diluted 1:64 times but not in one diluted 1:128 times, the titer of serum is 64. The more a sample can be diluted and still give a positive reaction, the greater the concentration of antibody in the sample and, hence, the higher the titer. The procedure for serially diluting a serum sample to determine titer is the same as serially diluting an antimicrobial agent to determine its minimum inhibitory concentration (MIC) for a bacterium as in Experiment 12.2.

Questions

1. Describe a doubling serial dilution of six tubes, beginning with a serum dilution of 1:2 in the first tube.

2. Define serum titer.

3. Why must both acute and convalescent sera be tested to make a serological diagnosis of infectious disease?

4. What is the difference between an agglutination test and a precipitation test?

5. In a paired serum sample, what test results indicate recent infection?

6. What is a humoral antibody?

7. Name the two types of antibodies that are produced following a microbial infection. What is the significance of each in serological diagnosis?

Principles and Practices of Infection Prevention

Learning Objectives

After completing this exercise, students should be able to:

1. Define nosocomial and health care-associated infections.
2. Name two gram-positive and two gram-negative bacteria that are important nosocomial pathogens.
3. Discuss the importance of Standard Precautions.
4. Explain the three types of Transmission-Based Precautions.
5. Discuss the basis of surveillance testing for epidemiologically significant bacteria.

General belief and public perception are that sick people who go to a hospital for medical and/or surgical care are cured of their illnesses and return home to their normal everyday activities. Although this scenario is true most of the time, some people who enter a hospital acquire or develop an infection during their hospital stay or soon after discharge. These hospital infections, called nosocomial, or hospital-associated, infections, are defined as any infection that is acquired 48 or more hours after admission and up to 48 hours after hospital discharge. Infections also may be acquired while a person is in a skilled nursing facility or home care setting, which has prompted the use of the term health care-associated infections (HAIs). Even so, HAIs occur most commonly in hospitals.

According to national studies, approximately 5% of hospitalized patients develop HAIs, with rates as low as 0.1% or as high as 20%, depending on the clinical setting. There are up to 2 to 4 million cases of HAIs per year and more than 90,000 deaths. In addition, nosocomial infections cause prolonged hospital stays, increased human suffering, and a dramatic increase in health care costs. By one estimate, HAIs account for 8 million additional patient hospital days per year, with a projected increased cost of 5 to 10 billion dollars.

Microbial Causes and Reservoirs of Infection

Nosocomial infections may be acquired *exogenously* from the hospital environment or *endogenously* from the patient's own normal flora. The most common HAIs involve the urinary tract (40%), surgical incisions (19%), the respiratory tract (15%), and the gastrointestinal tract (12%). Although any microorganism may cause HAI, gram-negative bacteria, usually from the gastrointestinal tract (*Escherichia coli* and *Klebsiella* and *Pseudomonas* spp.), are responsible for more than 50% of nosocomial infections. Gram-positive bacteria (staphylococci and streptococci) and yeasts (*Candida* spp.) account for most of the remainder.

The bacteria causing these infections are frequently acquired from the hospital environment. The sources are patient contact with contaminated inanimate objects (dishes, books, mops), known as fomites; or contact with health care personnel who might be carrying the organism on their hands or at another body site. A major concern is that these bacteria are likely to be resistant to many different groups of antibiotics; that is, they are multidrug resistant. Examples of multidrug-resistant gram-positive bacteria are coagulase-negative staphylococci; enterococci resistant to vancomycin, known as vancomycin-resistant enterococci or VRE; and methicillin-resistant *Staphylococcus aureus* (MRSA). Gram-negative bacteria also have acquired a wide variety of resistance mechanisms against antibiotics. A common example is their ability to produce beta-lactamases, enzymes that inactivate antibiotics in the penicillin, cephalosporin, and carbapenem classes. Gram-negative bacteria that produce beta-lactamases can be highly multidrug resistant, especially if they elaborate extended-spectrum beta-lactamase (ESBL) and carbapenemase. Either of these enzymes is sometimes produced by various members of the *Enterobacteriaceae,* such as *E. coli* and *Klebsiella* spp., as well as various members of the genus *Pseudomonas.* The infections caused by beta-lactamase-producing organisms are difficult to treat, because the bacteria are resistant to most antibiotics and few therapeutic options remain. The increased frequency with which multidrug-resistant bacteria are being encountered is one of the most serious public health problems today.

Guidelines for Control of Nosocomial Infections

To control the incidence and spread of nosocomial infections, infection control guidelines have been mandated by the Centers for Disease Control and Prevention and implemented in all hospitals. At first, these guidelines were disease specific, so patients with clearly identified infections were managed by the use of specific restrictions and techniques. Health care personnel tended to work with infected patients and their infectious materials with far greater care than they did with patients who were not identified as being infected. With the emergence of the HIV/AIDS epidemic in 1985, these infection control practices were reexamined and revised because of the potential that increased numbers of undiagnosed patients with HIV could be hospitalized. As a result, the Centers for Disease Control and Prevention established more stringent guidelines for handling all patients and their body substances.

Standard Precautions

Initially, these guidelines were called Universal Precautions, because they assumed that all patients and their specimens harbor infectious agents and must be treated with the same level of care. However, because most people with bloodborne viral infections, such as HIV or hepatitis B, do not have symptoms and cannot be visually recognized as infected, the term Standard Precautions has replaced the former designation of Universal Precautions. In brief, Standard Precautions are designed for the care of **all** persons (patients and health care staff), regardless of whether they are infected. The key components of Standard Precautions and their use are summarized in table 34.1.

Standard Precautions apply to blood and other body fluids, secretions and excretions (except sweat), nonintact skin, and mucous membranes. Their implementation is meant to reduce the risk of transmitting microorganisms from known or unknown sources of infection (e.g., patients, contaminated objects, and used needles and syringes) within the health care system. Applying Standard Precautions has become the primary strategy for preventing nosocomial infections in hospitalized patients.

Placing a physical, mechanical, or chemical barrier between microorganisms and an individual is highly effective for preventing the spread of infections. The barrier serves to break the disease transmission cycle.

Body Substance Isolation and Transmission-Based Precautions

At about the time that the guidelines for Standard Precautions were developed and introduced, another approach, called Body Substance Isolation (BSI), was devised. BSI focused on protecting patients and health care personnel from all moist and potentially infected body substances, including secretions and excretions, not just blood. BSI resulted in the development of three categories of Transmission-Based Precautions: Contact Precautions, Droplet Precautions, and Airborne Precautions (table 34.2). Contact Precautions are intended to prevent transmission of

Table 34.1   Standard Precautions: Key Components

Handwashing (or using an antiseptic hand rub)
- After touching blood, body fluids, secretions, excretions, and contaminated items
- Immediately after removing gloves
- Between patient contact

Gloves
- For contact with blood, body fluids, secretions, and contaminated items
- For contact with mucous membranes and nonintact skin

Masks, goggles, face masks
- Protect mucous membranes of eyes, nose, and mouth when contact with blood and body fluids is likely

Gowns
- Protect skin from blood or body fluid contact
- Prevent soiling of clothing during procedures that may involve contact with blood or body fluids

Linen
- Handle soiled linen to prevent touching skin or mucous membranes
- Do not pre-rinse soiled linens in patient care areas

Patient care equipment
- Handle soiled equipment in a manner to prevent contact with skin or mucous membranes and to prevent contamination of clothing or the environment
- Clean reusable equipment before reuse

Environmental cleaning
- Routinely care, clean, and disinfect equipment and furnishings in patient care areas

Sharps
- Avoid recapping used needles
- Avoid removing used needles from disposable syringes
- Avoid bending, breaking, or manipulating used needles by hand
- Place used sharps in puncture-resistant containers

Patient resuscitation
- Use mouthpieces, resuscitation bags, or other ventilation devices to avoid mouth-to-mouth resuscitation

Patient placement
- Place patients who contaminate the environment or cannot maintain appropriate hygiene in private rooms

infectious agents, including epidemiologically important microorganisms (e.g., MRSA, VRE, and *Clostridium difficile*), which are spread by direct or indirect contact with the patient or the patient's environment. Droplet Precautions are intended to prevent the transmission of pathogens, such as *Bordetella pertussis* and influenza virus, which are spread through close respiratory or mucous membrane contact with respiratory secretions. Typically, these respiratory pathogens do not remain infectious over a long distance in a health care facility, and therefore, no special air handling or ventilation is required in the patient's room. On the other hand, Airborne Precautions are intended to prevent transmission of agents, such as *Mycobacterium tuberculosis*, which remain infectious even over long distances when suspended in the air. The preferred placement for patients who require Airborne Precautions is in a negative-pressure room, also known as an airborne-infection isolation room. In this room, air is vented directly to the outside or is recirculated to the room through high-efficiency particulate air (HEPA) filtration before return. A summary of the Centers for Disease Control and Prevention guidelines for Standard Precautions and the three types of Transmission-Based Precautions are shown in table 34.2.

Principles and Practices of Infection Prevention

Table 34.2   Key Points in the Centers for Disease Control and Prevention Guideline for Isolation Precautions in Hospitals

Feature	Standard Precautions	Contact Precautions	Droplet Precautions	Airborne Precautions
Patient room	Standard	Private	Private	Private; door closed; well ventilated (minimum 6 air changes per hour; negative pressure)
Gloves	Before contact with blood, body fluids, mucous membranes, secretions, excretions, or broken skin	Before entering room	Standard	Standard
Handwashing	After glove removal; between patients	Standard; with antiseptic soap	Standard	Standard
Gown	Before procedure likely to generate projections of blood, body fluids, secretions, or excretions	Before contact with patient; if patient has diarrhea or open drainage of wounds or secretions	Standard	Standard
Mask	Before procedure likely to generate projections of blood, body fluids, secretions, or excretions	Standard	If within 3 feet of patient	Before entering room
Examples of conditions/patients that precautions apply to	All patients	• Multidrug-resistant bacteria of special clinical and epidemiologic significance (e.g., MRSA, VRE) • Major abscess, cellulitis, or decubiti • *Clostridium difficile* infection • Acute diarrhea in an incontinent or diapered patient • RSV infections, bronchiolitis, and croup in young infants	• Meningitis • Diphtheria • Pertussis • Influenza • Mumps • Rubella • Streptococcal pharyngitis, pneumonia, or scarlet fever in young children	• Tuberculosis or suspected tuberculosis • Measles • Varicella, disseminated zoster

Surveillance Testing

Occasionally, patients, health care workers, and, to a lesser extent, the hospital environment are tested to determine if they are colonized with microorganisms that may serve as reservoirs for transmission of nosocomial infections. MRSA is one organism that may be screened for regularly in patients and health care personnel. Typically, surveillance samples are collected from anatomic sites that are known to have high colonization rates with *S. aureus*. These anatomic sites include the anterior nares, axillae, and inguinal (groin) area. The samples are then submitted to the microbiology laboratory, where they are tested for the presence of MRSA, either by culture or by a molecular PCR test that screens for the *S. aureus mecA* gene, which is responsible for methicillin resistance. If MRSA is detected by either method, the patient is placed in a special isolation room to prevent MRSA transmission to other patients and staff. In addition, the patient may also be treated to eradicate the MRSA carrier state. The use of MRSA screening tests has become more widespread in hospitals throughout the United States. Many hospitals routinely screen patients for MRSA colonization before they undergo heart bypass surgery or prosthetic joint replacement. To minimize the risk of MRSA transmission to other patients, some hospitals screen all patients before admission. Surveillance testing can also be performed to determine if patients are colonized with VRE or bacteria that produce ESBL or carbapenemases.

In addition to performing surveillance cultures on patients, it is often necessary for the laboratory to determine the sterility of other substances, such as dialysis fluids, reagent water, and blood components, and to perform sterility tests to document proper use of aseptic technique by employees who prepare intravenous fluids. The laboratory must document the procedures used for such assays and retain laboratory test results for specific time periods.

A specialized medium, called CHROMagar MRSA, has been developed to identify patients who might be colonized with MRSA. CHROMagar MRSA is a selective and differential medium that permits the detection and identification of methicillin-resistant *Staphylococcus aureus* (MRSA) colonies directly from the patient specimen. The medium contains specific chromogenic substrates and the antibiotic cefoxitin. MRSA strains grow in the presence of cefoxitin and produce mauve-colored colonies, resulting from the hydrolysis of the chromogenic substrate. Methicillin-susceptible *Staphylococcus aureus* (MSSA) strains are not able to grow on this medium, because they are inhibited by cefoxitin. Additional selective agents are incorporated to suppress growth of gram-negative bacteria, yeast, and some gram-positive cocci. Some bacteria other than MRSA may grow on this medium, but if they use the chromogenic substrates, they produce blue/green-colored colonies while other non-MRSA bacteria that cannot use the substrates produce white or colorless colonies.

EXPERIMENT 34.1   Nasal Surveillance Cultures for MRSA

In this experiment, you will perform a nasal surveillance screening test for MRSA.

Purpose	To culture a nasal specimen for the presence of MRSA
Materials	A simulated or student-collected nasal specimen
	Sheep blood agar plate (BAP)
	CHROMagar MRSA medium
	Wire or disposable inoculating loop
	Marking pen or pencil
	Incubator

Principles and Practices of Infection Prevention

Procedures

1. A simulated nasal specimen will be provided by your instructor **or** your instructor may have students collect an anterior nares specimen on themselves or a laboratory partner. An anterior nares specimen is collected by inserting a sterile cotton-tipped swab into the anterior nares and rotating the swab with gentle pressure against the inside surface of the nares for 5 to 10 seconds.
2. Inoculate the nasal swab specimen onto BAP and CHROMagar MRSA media by rubbing the surfaces of the swab across an area of approximately ¼ of the plate.
3. Streak the plates to obtain isolated colonies by using a heat-sterilized wire loop or sterile, disposable plastic loop. Label the plates with the "patient's" name and place them in a 35°C incubator.
4. After 18 to 24 hours of incubation, examine the BAP and CHROMagar plates for growth.

Results

1. Observe the appearance of the colonies and the amount of growth on each medium.
2. Record your observations in the following table.

Name of Medium	Number of Colonies	Types of Colony Morphology	Color of Colony	Culture + or − for MRSA*
Blood agar plate				
CHROMagar MRSA				

*If materials are available, students may perform a staphylococcal latex agglutination test on suspected *S. aureus* colonies and a rapid latex or immunochromatographic assay to detect penicillin-binding protein 2a (PBP2a), a marker for *mecA*, the methicillin resistance gene.

Questions

1. What is the definition of a nosocomial or health care-associated infection?

2. Name three anatomic sites that are the most common sources of nosocomial infection. Do these infections result from an exogenous or endogenous source?

3. How did the HIV epidemic result in the development of Universal Precautions?

4. Why are Standard Precautions applied to the care of all patients?

5. What are the differences between: Contact Precautions, Droplet Precautions, and Airborne Precautions?

6. For each of the three types of Transmission-Based Precautions, give two examples of an infection to which they apply.

7. Why is MRSA screening of patients important?

8. Name three anatomic sites that are often screened for the presence of MRSA.

9. What other types of sterility testing may the laboratory be called upon to perform?

SELECTED LITERATURE AND SOURCES

American Hospital Formulary Service. 2017. *AHFS 2017 drug information.* Bethesda, MD: American Society of Health-System Pharmacists.

Anaissie, E. J., McGinnis, M. R., and Pfaller, M. (eds.). 2009. *Clinical mycology.* 2nd ed. Philadelphia: Elsevier Health Sciences (includes CD-ROM).

Anderson, D., Salm, S., and Allen, D. 2015. *Nester's Microbiology: A human perspective.* 8th ed. New York: McGraw-Hill.

Ash, L. R., and Orihel, T. C. 2007. *Atlas of human parasitology.* 5th ed. Chicago: American Society of Clinical Pathology.

asmscience.org/content/education/imagegalleries

Bauerfeind, R., von Graevenitz, A., Kimmig, P., Scheifer, H. G., Schwarz, T., Slenczka, W., and Zahner, H. 2016. *Zoonoses: Infectious diseases transmissible from animals to humans.* 4th ed. Washington, DC: ASM Press.

Bennett, J. E., Dolin, R., and Blaser, M. J. 2014. *Mandell, Douglas, and Bennett's principles and practice of infectious diseases.* 8th ed. Philadelphia: Saunders.

Black, J. 2015. *Microbiology: Principles and explorations.* 9th ed. New York: Wiley.

Bogitsh, B. J., Carter, C. E., and Oeltman, T. M. 2012. *Human parasitology.* 4th ed. Waltham, MA: Academic Press.

Bollet, A. J. 2004. *Plagues and poxes: The impact of human history on epidemic disease.* 2nd ed. New York: Demos Medical Publishing.

Boose, J., and August, M. 2013. *To catch a virus.* Washington, DC: ASM Press.

Bottone, E. J. 2004. *Atlas of clinical microbiology of infectious diseases,* vol. 1, *Bacterial agents.* London: Parthenon.

Bottone, E. J. 2006. *Atlas of clinical microbiology of infectious diseases,* vol. 2, *Viral, fungal, and parasitic agents.* Oxford, UK: Taylor & Francis.

Centers for Disease Control and Prevention. 2016. *Healthcare Infection Control Practices Advisory Committee (HICPAC) general guidelines.* http/www.cdc.gov/hicpac/pubs.html

Chapel, H., Haeney, M., Misbah, S., and Snowden, N. 2014. *Essentials of clinical immunology.* 6th ed. Hoboken, NJ: Wiley-Blackwell.

CLSI. 2016. *Performance standards for antimicrobial susceptibility testing: 26th informational supplement, M100-S26.* Wayne, PA: CLSI.

Cowan, M. K. 2014. *Microbiology: A systems approach.* 4th ed. New York: McGraw-Hill.

de la Maza, L. M., Peterson, E.M., Pezzlo, M. T., Shigei, J. T., and Tan, G. L. 2013. *Color atlas of medical bacteriology.* 2nd ed. Washington, DC: ASM Press.

Detrick, B., Hamilton, R. G., and Schmitz, J. L. 2016. *Manual of molecular & clinical laboratory immunology.* 8th ed. Washington, DC: ASM Press.

Dixon, B. 2009. *Animalcules: The activities, impacts, and investigators of microbes.* Washington, DC: ASM Press.

Doan, T., Melvold, R., Viselli, S., and Waltenbaugh, C. 2012. *Immunology.* 2nd ed. Philadelphia: Lippincott Williams & Wilkins.

Doyle, M. P., and Buchanan, R. L. (eds.). 2013. *Food microbiology: Fundamentals and frontiers.* 4th ed. Washington, DC: ASM Press.

Echenberg, M. 2007. *Plague ports: The global urban impact of bubonic plague.* New York: New York University Press.

Engelkirk, P. G., and Duben-Engelkirk, J. 2014. *Burton's microbiology for the health sciences.* 10th ed. Philadelphia: Lippincott Williams & Wilkins.

Garcia, L. S. 2016. *Diagnostic medical parasitology.* 6th ed. Washington, DC: ASM Press.

Garcia, L. S. 2009. *Practical guide to parasitology.* 2nd ed. Washington, DC: ASM Press.

Gaynes, R. P. 2011. *Germ theory: Medical pioneers in infectious diseases.* Washington, DC: ASM Press.

Goering, R., Dockrell, H., Zuckerman, M., Roitt, I., and Chiodini, P. L. 2013. *Mim's medical microbiology.* 5th ed. Philadephia: Saunders.

Grabenstein, J. D. 2012. *Immunofacts 2013: Vaccines and immunologic drugs.* Philadelphia: Lippincott Williams & Wilkins.

Harvey, R. A., Cornelissen, C. N., and Fisher, B. D. 2012. *Microbiology.* 3rd ed. Philadelphia: Lippincott Williams & Wilkins.

Hayden, R. T., Wolk, D. M., Carroll, K. C., and Tang, Y.-W. 2016. *Diagnostic microbiology of the immunocompromised host.* 2nd ed. Washington, DC: ASM Press.

Heymann, D. L. (ed.). 2014. *Control of communicable diseases manual.* 20th ed. Washington, DC: American Public Health Association.

Jaykus, L-A., Wang, H. H., and Schlesinger, L. S. 2009. *Food-borne microbes: Shaping the host ecosystem.* Washington, DC: ASM Press.

Jorgensen, J. H. and Pfaller, M. A. (eds.-in-chief). 2015. *Manual of clinical microbiology.* 11th ed. Washington, DC: ASM Press.

Kauffman, C. A., Pappas, P. G., Sobel, J. D., and Dismukes, W. E. (eds.). 2011. *Essentials of clinical mycology.* 2nd ed. New York: Springer.

Kolter, R., and Maloy, W. 2012. *Microbes and evolution: The world that Darwin never saw.* Washington, DC: ASM Press.

Larone, D. H. 2011. *Medically important fungi. A guide to identification.* 5th ed. Washington, DC: ASM Press.

Leber, Amy L. (ed.). 2016. *Clinical microbiology procedures handbook.* 4th ed. Washington, DC: ASM Press.

Leboffe, M. J., and Pierce, B. E. 2011. *A photographic atlas for the microbiology laboratory.* 4th ed. Englewood, CO: Morton Publishing Co.

Loeffelholz, M. J. (ed.-in-chief). 2016. *Clinical virology manual.* 5th ed. Washington, DC: ASM Press.

Loker, E. S., and Hofkin, B. 2015. *Parasitology: A conceptual approach.* New York: Garland Science.

Madigan, M. T., Martinko, J. M., Bender, K. S., Buckley, D. H., Stahl, D. A., and Brock, T. 2014. *Brock biology of microorganisms.* 14th ed. New York: Pearson.

Mahon, C. E., Lehman, D. C., and Manuselis, G., Jr. 2014. *Textbook of diagnostic microbiology.* 5th ed. Philadelphia: Saunders.

Male, D., Brostoff, J., Roth, D., and Roitt, I. 2013. *Immunology*. 8th ed. Philadephia: Saunders.

McDonnell, G. E. 2007. *Antisepsis, disinfection, and sterilization: Types, action, and resistance*. Washington, DC: ASM Press.

The Medical Letter. 2015. *Handbook of antimicrobial therapy*. 20th ed. New Rochelle, NY: The Medical Letter, Inc.

Montville, T. J., Matthews, K. R., and Kniel, K. E. 2012. *Food microbiology: An introduction*. 3rd ed. Washington, DC: ASM Press.

Murphy, K. M., and Weaver, C. 2016. *Janeway's immunobiology: The immune system*. 9th ed. New York: Garland Science.

Murray, P. R., Rosenthal, K. S., and Pfaller, M. A. 2015. *Medical microbiology*. 8th ed. New York: Elsevier.

Pagana, K. D., Pagana, T. J., and Pagana, T. N. 2016. *Mosby's diagnostic and laboratory test reference*. 13th ed. St. Louis, MO: Mosby.

Persing, D. (ed.-in-chief). 2016. *Molecular microbiology: Diagnostic principles and practice*. 3rd ed. Washington, DC: ASM Press.

Pommerville, J. 2015. *Alcamo's fundamentals of microbiology*. 10th ed. Sudbury, MA: Jones and Bartlett.

Procop, K., and Koneman, E. 2016. *Koneman's color atlas and textbook of diagnostic microbiology*. 7th ed. Philadelphia: LWW.

Rich, R. R., Fleisher, T. A., Shearer, W. T., Schroeder, H., Frew, A. J., and Weyand, C. N. 2013. *Clinical immunology*. 4th ed. Philadelphia: Saunders.

Richman, D., Whiteley, R. J., and Hayden F. G. 2016. *Clinical virology*. 4th ed. Washington, DC: ASM Press.

Scheld, W. M., Hughes, J. M., and Whitley, R. J. 2016. *Emerging infection 10*. Washington, DC: ASM Press.

Schlossberg, D. (ed.). 2011. *Tuberculosis and nontuberculous mycobacterial infections*. 6th ed. Washington, DC: ASM Press.

Sherman, I. W. 2007. *Twelve diseases that changed the world*. Washington, DC: ASM Press.

Siegel, J. D., Rhinehart, E., Jackson M., and Chiarello, L. 2007. *Guideline for isolation precautions: Preventing transmission of infectious agents in healthcare settings*. http://www.cdc.gov/ncidod/dhgp/pdf/isolation2007.pdf.

Sompayrac, L. M. 2015. *How the immune system works*. 5th ed. Boston: Wiley-Blackwell.

Swanson, M., Reguera, G., Schaecter, M., and Neidhardt, F. 2016. *Microbe*. 2nd ed. Washington, DC: ASM Press.

Talaro, K. P., and Chess, B. 2014. *Foundations in microbiology*. 9th ed. New York: McGraw-Hill.

Tille, P. 2016. *Bailey & Scott's diagnostic microbiology*. 14th ed. St. Louis, MO: Mosby.

Todar, K. *Todar's online textbook of bacteriology*. http://textbookofbacteriology.net.

Tortora, G. J., Funke, B. R., and Case, C. L. 2015. *Microbiology: An introduction*. 12th ed. Upper Saddle River, NJ: Pearson Education.

Turgeon, M. E. 2017. *Immunology & serology in laboratory medicine*. 6th ed. St. Louis, MO: Mosby.

Wilcox, J. B. 2012. *Hospital acquired infections*. Hauppauge, NY: Nova Science Publishers.

Willey, J. M., Sherwood, L. M., and Woolverton, C. J. 2016. *Prescott's microbiology*. 10th ed. New York: McGraw-Hill.

www. microbelibrary.org. Washington, DC: American Society for Microbiology.

Zeibig, E. A. 2012. *Clinical parasitology, a practical approach*. 2nd ed. Philadelphia: Saunders.

INDEX

Note: Page numbers followed by f refer to illustrations; page numbers followed by t refer to tables.